Epic Disruptions

Epic Disruptions

11 Innovations That Shaped Our Modern World

Scott D. Anthony

Harvard Business Review Press
Boston, Massachusetts

> **HBR Press Quantity Sales Discounts**
>
> Harvard Business Review Press titles are available at significant quantity discounts when purchased in bulk for client gifts, sales promotions, and premiums. Special editions, including books with corporate logos, customized covers, and letters from the company or CEO printed in the front matter, as well as excerpts of existing books, can also be created in large quantities for special needs.
>
> For details and discount information for both print and ebook formats, contact booksales@harvardbusiness.org, tel. 800-988-0886, or www.hbr.org/bulksales.

Copyright 2025 Scott D. Anthony

All rights reserved

Printed in the United States of America

10 9 8 7 6 5 4 3 2 1

No part of this publication may be reproduced, stored in or introduced into a retrieval system, or transmitted, in any form, or by any means (electronic, mechanical, photocopying, recording, or otherwise), without the prior permission of the publisher. Requests for permission should be directed to permissions@harvardbusiness.org, or mailed to Permissions, Harvard Business School Publishing, 60 Harvard Way, Boston, Massachusetts 02163.

The web addresses referenced in this book were live and correct at the time of the book's publication but may be subject to change.

Library of Congress Cataloging-in-Publication Data

Names: Anthony, Scott D. author
Title: Epic disruptions : 11 innovations that shaped our modern world / Scott D. Anthony.
Description: Boston, Massachusetts : Harvard Business Review Press, [2025] | Includes index. |
Identifiers: LCCN 2025001516 (print) | LCCN 2025001517 (ebook) | ISBN 9781647829711 hardcover | ISBN 9781647829728 epub
Subjects: LCSH: Technology—Social aspects—History | Disruptive technologies—History | Technological innovations—History
Classification: LCC T14.5 .A575 2025 (print) | LCC T14.5 (ebook) | DDC 303.48/309—dc23/eng/20250606
LC record available at https://lccn.loc.gov/2025001516
LC ebook record available at https://lccn.loc.gov/2025001517

ISBN: 978-1-64782-971-1
eISBN: 978-1-64782-972-8

The paper used in this publication meets the requirements of the American National Standard for Permanence of Paper for Publications and Documents in Libraries and Archives Z39.48-1992.

Contents

Introduction   1
The Power of Disruption

1 The Big Bang   13
Gunpowder and Its Explosive Impact

2 The First Information Revolution   29
How the Printing Press Changed Everything

3 Bacon (Not the Food) and Boyle   49
Laying the Foundations of Modern Science

4 Florence from Florence   67
Revolutionizing Health Care with Images and Words

5 The Great Democratization   85
How the Model T Put the World on Wheels

6 Another Big Bang   103
The Transistor's Unexpected Path to Transformation

7 The Recipe for Disruption   123
Julia Child and the Art of French Cooking

8 The Special Sauce — 143
McDonald's Fast-Food Breakthrough

9 Oh Crap! — 159
Pampers Disposes of Its Competitors

10 The Ghosts of Bethlehem — 179
Big Steel and the Shaping of an Industry

11 Anomalies Wanted — 201
The iPhone and the Era of Smart Innovation

Conclusion — 221
What's Next? The Future of Disruption

Notes	239
Index	251
Acknowledgments	263
About the Author	267

Epic Disruptions

Introduction

The Power of Disruption

Disruption.

Say the word out loud.

Dis-rup-tion.

Did your lip curl? Your mouth purse? Did you spit the three syllables out?

What thoughts went through your mind? Perhaps something that interrupted something you were working on. Maybe you remembered a bad experience at an airport. I'm willing to bet that whatever it was, it was a negative feeling.

When I see the word, my mind goes somewhere else. I flip *disruption* from a noun to the adjective *disruptive*. I then put that adjective before the word *innovation*.

Disruptive innovation.

I think of the disruptive innovations that surround me. I am typing on a Lenovo ThinkPad X1 powered by an Intel vPro processor. Bell Labs' invention of the transistor ultimately brought computing out of centralized labs in corporations and governments in to people's homes. A message just appeared on my iPhone. Smartphones and social networking have brought communication to new contexts. I just enjoyed a homemade espresso that rivaled what you could get in a café in Italy. Disruptive innovations like Nestlé's Nespresso machine have brought to millions of people what was once available only in inaccessible locations. I live in the suburbs of Boston. The very nature of my life is shaped by Henry Ford's breakthrough Model T, which ushered in the automotive age. I see an Amazon delivery truck out the window. Just think, a generation ago I had to (*shudder*) go to a store or (*bigger* shudder) *wait* to get something I needed. If you are reading a printed version of this book, thank the printing press, which democratized knowledge itself. Maybe you were one of seventy million people who ate at McDonald's today, the result of a disruptive model honed decades ago.

Disruption is an engine of progress. By making the complicated simple and the expensive affordable, it transforms how we work, play, live, and communicate.

I then think about how disruptive innovations have dramatically changed industries, serving as a buzz saw felling great companies such as Digital Equipment Corporation, Nokia, Sears, Eastman Kodak, Blockbuster, numerous newspaper companies, and many more.

I think of companies today wrestling with disruptive change. There are major automakers like General Motors and Volkswagen confronting the electrification of passenger cars, autonomous vehicles, and ride sharing services like Uber. Major energy companies like ExxonMobil and Shell are grappling with dramatic changes in the renewable energy market. And countless companies are placing bets on how generative artificial intelligence (gen AI) will change their businesses.

Disruption is a driver of industry change. It creates tomorrow's great companies and causes today's great companies to fail.

I think next about how disruptive innovation continues to reshape our world. Technologies are advancing at an exponential rate. The lines between industries are blurring. The expectations of consumers, colleagues, and children are shifting rapidly. And a complex world throws shock after shock at us.

Disruption is ever present.

And finally, I think about Clayton "Clay" Christensen, the man who coined the term *disruptive innovation* and shaped how I understand the world. My teacher. My mentor. My colleague. My friend. He taught me to see disruption not just as upheaval but also as opportunity—an unstoppable force that demands our attention.

And that's what I see in the word *disruption*.

A Brief Overview of Disruptive Theory

I have been obsessed with disruptive innovation since I met Clay in 2000.* I was a second-year student at Harvard Business School (HBS).

I took the first version of a class called Building and Sustaining a Successful Enterprise, which Clay created. Clay was enough of a name that even though there was no course history, he filled two 80-person sections of the class. A total of 7,648 students took the class between 2000 and 2018. At its peak, about two-thirds of the 900 HBS students would take a version of this class taught by Clay and a team of teachers.[1]

I read Clay's book *The Innovator's Dilemma* on October 21, 2000, on a flight to Phoenix to see Pearl Jam with my friend Chris Kao.** The book resonated. Deeply. Clay started the research recounted in

---

* A note on names: Most of the time, I will use last names on subsequent mentions. Sometimes I will use first names if doing so seems to make more sense in the writing. And hello there, footnote readers. Footnotes are for asides like this. All references appear in the endnotes.

** That was to be my fourth Pearl Jam concert. On September 17, 2024, I saw my twenty-fourth in Fenway Park in Boston. I'm a fan.

the book with a powerful question: What causes seemingly smart, well-run companies to fail?

It's a doozy of a question. Understanding what causes stupid, poorly run companies to fail is obvious. But smart companies? Those run by competent managers? That are doing things that *feel* right? Those failures are interesting. No, fascinating.

Clay's question led him to study the disk drive industry. The industry was young, essentially created in 1965, when IBM allowed plug-compatible disk drives into its System 360. And the industry changed quickly.

I remember Clay saying that former HBS Dean Kim Clark had told him that the disk drive industry was the "fruit fly" of the business world. When I relayed this to Kim, he laughed and said he didn't remember saying exactly that, but he added, "I encouraged him to look at industries that move really fast, because you'll have lots of technological transitions."[2]

Clay found that market leaders in the disk drive industry tended to win when a technology offered improvement along traditional dimensions and tended to lose when an innovation disrupted and redefined what performance meant. As he wrote in a 1993 article in *Business History Review*, "Disruptive architectures did not necessarily represent a sequence of successfully inferior technological approaches—they simply offered a very different package of attributes. Trajectory-disrupting architectures therefore tended to be used initially in new, emerging market segments rather than in the large, mainstream markets served by the leading disk drive manufacturers."[3]

A 1995 *Harvard Business Review* article by Clay and his thesis adviser Joe Bower introduced a simple model to capture this idea (figure I-1). The model has performance on the vertical axis and time on the horizontal axis. The model has two differently sloped trajectories. The relatively flat line shows the performance that a given group of customers demands. It represents the problem they are seeking to solve, the job

Figure I-1

The original disruptive innovation line model

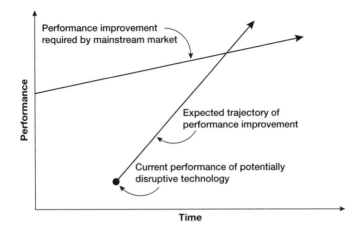

Source: Joseph L. Bower and Clayton M. Christensen, "Disruptive Technologies: Catching the Wave," *Harvard Business Review*, January–February 1995.

they are trying to get done. The steeper line shows the performance that a company provides. It is steeper because Clay argued that *companies innovate faster than people's lives change.*

That simple statement has profound implications. In the early days, an innovation might not be good enough for a given group of customers. As companies fire up their innovation engines, that which once was not good enough becomes perfectly adequate and then becomes *too* good. Clay called this *overshooting*.

Why do companies overshoot their customers? In business, gravity pulls you up, not down. Companies generally seek to solve the hard problems, to serve the most demanding customers, because that's the path to profits. People will always take improvements but will grow increasingly unwilling to pay for them, creating opportunities for people to change the game.

Clay's research highlighted two types of innovations: those that *sustain* an established trajectory and those that *disrupt* by trading off performance along dimensions that historically mattered to mainstream customers through improved performance on different dimensions.* In other words, the disruption is *less good* in some ways, typically in terms of raw performance, and *better* in historically overlooked dimensions such as simplicity and affordability.

Drawing on research Bower conducted in the 1970s, Clay argued that companies tend to allocate resources toward opportunities that allow them to offer better performance to their best customers, because historically such investments have been rewarded with premium prices, high margins, and so on. Mainstream customers don't want inferior products, at least initially. So even if a company spots an opportunity to disrupt, its customers don't want it to do so.

This is the innovator's dilemma. You do what you are supposed to do. Listen to your best customers. Innovate to meet their needs. Push prices up. Push margins up. And fail in the face of disruptive change.

That's about it. Reasonably straightforward. Deeply profound.

After I took Clay's class, I spent two years leading up his research team, and then joined the team at Innosight, a consulting company he cofounded to help companies use his research. I played various roles at Innosight over the next two decades, including a stint as its global leader from 2012 to 2018. I have advised companies around the world about disruption, helped to build a company grounded in the idea of disruption, invested in startup companies that had potential to drive disruptive growth, and even tried to incubate a few additional disruptive ideas. In 2022 I started teaching students about disruption at the Tuck School of Business at Dartmouth College. In short, I've been obsessed with disruption for more than two decades.

---

* He would play with this simple model over time. At one point, he created two forms of disruption: low-end and new-market disruptions, with new-market disruption depicted on a third dimension. Later, he would talk about the difference between efficiency, sustaining, and disruptive innovations. He called disruptive innovations *market-creating innovations* in his last book, *The Prosperity Paradox*. While Clay was dogmatic in some areas, he was fluid in others.

I define a disruptive innovation as one that transforms an existing market or creates a new one by making the complicated simple and the expensive affordable. The chapters ahead will describe disruptive innovations in much more depth, but here are two brief examples to bring the idea to life.

Photography was invented in the early eighteenth century. Producing a picture was technologically complex. It required an individual to learn how to mix chemicals and manage a very delicate process that carried personal risk. What allowed photography to become a mass-market phenomenon was George Eastman's Kodak Brownie box camera. The device was technology simple and affordable. But even more critically, Eastman separated *capturing* pictures and *developing* pictures.

The tagline: "You press the button, we do the rest." Users mailed the camera to Kodak and received pictures in the mail. It made the complicated simple and set Kodak up for more than a century of profitable growth ... until disruption struck again (more on that in chapter 2).

Next, consider Alexander Graham Bell's telephone. Bell wasn't trying to unseat the communications giant of his time—Western Union, which offered telegraphy service. He was trying to *help* Western Union. The patent for the telephone was titled "Improvements in Telegraphy." Bell offered to sell the technology to Western Union for a pittance of $100,000 ($3 million in modern terms).* Western Union passed. Its president, William Orton, said, "What use could this company make of an electrical toy?"[4]

In hindsight, this was a bad decision, but at the time, it made sense. The technology offered new benefits—the ability to transmit a human voice over a wire using electricity—but it was too limited for Western

---

* Inflation-adjusted figures in this book are based on Ian Webster's Inflation Calculator at https://www.in2013dollars.com/us/inflation. Why that site? Google recommended it!

Union to offer better service to the banks and railroads that relied on the telegraph to coordinate transactions and scheduling.

So, Bell created a new company and found consumers who were thrilled with his limited technology, namely, wealthy consumers looking to communicate inside their own homes. The telephone had its limitations, but it was better than yelling or running through the house. The company that commercialized Bell's technology ultimately became AT&T. Its research arm birthed the transistor (discussed in chapter 6).

Clay Christensen's original work focused on why great companies fail. The more he studied and worked with the phenomena, the more he shifted his sights to the power of disruption to transform entire industries and create new opportunities. Disruption can create dilemmas. It can also create opportunities.

And his ideas spread. Intel successfully launched a stripped-down, simplified processor called Celeron to protect itself from a potentially disruptive attack. In 1999, *Forbes* ran a cover story titled "Andy Grove's Big Thinker: Clayton Christensen Tells How to Survive Disruptive Technologies." The cover photo had Clay's right arm over Grove's left shoulder. The image reinforced the word "Big" in the title, as, at six feet eight, Clay was about a foot taller than the Intel CEO.

Then the idea exploded. The concept of disruptive innovation seemed to tie so nicely into what was going on in the world economy in the first decade of the 2000s (the aughts) with the rise of companies like Amazon, Meta, and Google and the decline of traditional department stores, newspapers, and the yellow pages.*

---

* Get this, kids. In every home, there used to be a book, a yellow book that provided a categorized, alphabetical listing of local businesses with their phone numbers. It was thick. In some cases, so thick it had to be split into two volumes. This was a $13 billion industry in 2000. That's billion with a *b*. It's largely gone now, of course, so there's a less pressing need for companies to name themselves "Aaalan's Pest Removal" to be in the running for that coveted first listing.

And the idea continues to fascinate. It's alive and accelerating in today's world, fueled by rapid technological advances (AI, anyone?) and shifting consumer expectations. Understanding the patterns of disruption is essential—not just for businesses hoping to survive it, but for anyone else striving to thrive in an ever-changing world.

As you turn the pages ahead, you'll see how disruptive innovations have repeatedly reshaped our modern world. From gunpowder to Julia Child's *Mastering the Art of French Cooking*, the transistor, and Big Macs, these innovations didn't start by perfecting the old—they created new value in overlooked ways.

So, as you read on, I invite you to see disruption through the lens Clay shared with me. It isn't just about the fall of the old. It's about the rise of what comes next.

Four Unanswered Questions

So, why this book, then?

For all its power, disruption is still widely misused and misunderstood. As the saying goes, history may not repeat, but it often rhymes.*

So, this book represents a romp through history to see what we can learn from historical case studies of disruption. Now, I'm neither a historian nor a trained academic, so let's call this not a history book but a *history-ish* book. I have written enough business books to know that every story is, well, a story, with abstractions and simplifications in the name of helping to communicate an underlying concept clearly. Business books are all works of fiction. This book is, too, though one grounded in research with the endnotes to prove it.

---

* Like many of you, I suspect, I attributed that line to Mark Twain. Eagle-eyed copyeditor Patricia Boyd said there's no proof Twain ever said it. The first published comment of this nature was by psychoanalyst Theodor Reik. And now we all know!

My previous books were case-study-driven framework books. They provided practical tools to improve the reader's ability to innovate and grow. The case studies illustrated the tools and built confidence in their validity. This book is *not* such a book. As hard as it was for me, I sought to avoid the tendency to cram in tools, frameworks, and checklists. Models appear when they are a part of the story or bring unique insight into the story. The case studies are essentially in chronological order rather than grouped by topics.

I entered the research with four questions. Let me summarize the questions and tell you the implications I drew from the stories. You might draw very different conclusions from these stories. It's all good.

Who does it? And I don't mean what type of company does it. Is disruptive innovation about a lone genius? Or is it something done by groups and teams? From the very first story of gunpowder through to the ending story of the iPhone, it seems clear to me that disruption is a *collectively individualistic* activity. It isn't a faceless act done by an amorphous organization; it is an intensely human activity. Of course, there are those who play outsize roles, but ultimate disruptive success has many hands, with handoffs that typically stretch over decades if not centuries.

Is it random? If you look at some of the academic criticism about the idea of disruptive innovation, there's a theme that disruption is obvious in hindsight but not clear in the moment. That observation prompts two big questions: Can you predict success? And can you take actions to improve the probability of success? My conclusion is that disruption is *predictably unpredictable*. There are clear patterns that connect the stories in this book, but there are also apparently random moments. That leads me to believe you can tilt the odds in your favor. But this isn't, and really can't be, Newtonian physics.

Is it accelerating? The story of gunpowder stretches across centuries. The iPhone, on the other hand, seems to be an overnight success story. It *feels* as if the pace of disruption is accelerating, perhaps even exponentially. Is it? No doubt, the world is moving faster than ever before. At the same time, however, if you look deeply at any story, you will see roots that extend surprisingly far into the past. Success, therefore, requires *patient perseverance* over years, and sometimes decades.

Is it a universal good? I gave ChatGPT the following words: children, death, growth, love, peace, taxes, war, unicorns . . . and innovation. I asked it, based on its training, to score each word on a scale of zero being negative and ten being positive. Innovation scored an eight, tied with children and growth, and a smidge behind love and unicorns. That feels right, doesn't it?* Innovation is a positive thing. Everyone thinks that, right? But I've seen the downside of disruption, the struggles companies have, the dark side to it. And, indeed, looking deep into history, you see clearly that disruption casts a shadow.

I didn't write this book linearly, so you don't have to read it linearly.** Table I-1 shows which of the four questions each chapter focuses on the most.

I made a conscious choice to stop with the 2007 launch of the iPhone, to focus on stories that are generally complete, ones whose implications are clear. What to say about emerging disruptions like AI and additive manufacturing?

---

  * Well, as much as I really do love my children, some days it is a bit hard to see that eight-point score!
 ** For those of you interested in the process of writing a book, version 1 of the manuscript had a full-length chapter about Clay and his research near the end of the book, version 2 redid that chapter and made it the book's first chapter, version 3 blew it up and incorporated it as part of this introduction. If you are curious about what version 2 looked like, shoot me an email at scott.d.anthony@tuck.dartmouth.edu.

Table I-1

Overview of case studies in *Epic Disruptions*

Chapter and Case Study	Who Does It?	Is It Random?	Is It Accelerating?	Is It a Universal Good?
1. Gunpowder	X		X	
2. The printing press		X		X
3. The scientific revolution		X		X
4. Florence Nightingale	X			X
5. The Model T	X			X
6. The transistor		X	X	
7. Julia Child	X	X		
8. McDonald's	X	X		
9. Disposable diapers		X	X	
10. Steel minimills				X
11. The Apple iPhone	X	X	X	X

I address this question in the conclusion, which rounds out the book by considering how Clay would make sense of ongoing developments that might have significant disruptive potential. Spoiler alert: The conclusion provides no answers. Rather, it provides a way to think through what is going on now.

Once I started describing the book to people, the whataboutisms would start. "What about fertilizer? You have to have fertilizer!" one friend said. "What about food processing?" asked another. Country X. Industry Y. Everyone has their favorite. I picked mine.* I sought an interesting and diverse variety of stories for readability and to create the space to explore a wide range of ideas.

So, let the wild rumpus start.** With a bang.

---

* If you are curious, and by reading this footnote, you have displayed your curiosity, the final two cuts were venture capital (which ended up being folded into chapter 6) and cleantech solutions (which felt too in-progress to work in parallel to the other chapters).

** I have four kids. I've read *Where the Wild Things Are* at least a hundred times.

1

The Big Bang

Gunpowder and Its Explosive Impact

The Theodosian Walls of Constantinople, the Queen of Cities, the bridge between East and West, were the stuff of legend. Constantine I started construction after his victory in the Battle of Chrysopolis in 324. When completed during the reign of Theodosius II, the walls provided multiple layers of defense. Attackers first encountered a fifteen-foot-deep, sixty-foot-wide moat. If they survived arrows hailing down on them while crossing the moat, they next encountered a twenty-five-foot-tall limestone and brick wall. There was a small stretch of cleared ground and then an impossibly tall wall, forty feet high and fifteen feet thick, buttressed by ninety-six towers.

As legendary as they were, these walls weren't perfect.

Picture Justinian II contemplating those walls in 705. Perhaps he stroked his golden nose, relishing his coming redemption. What's the deal with that golden nose, you wonder? The son of Constantine IV became emperor at the tender age of sixteen in 685. A brutal ruler, he lost the support of both peasants and aristocrats. His mutilation after being deposed earned him memorable monikers such as Justinian No-Nose and Justinian Slit-Nose. In 705, he brought fifteen thousand troops to the edge of the city. He couldn't go *over* the walls. He couldn't break them down. But he could go *under* them. His troops sneaked into Constantinople though a secret sewer and quickly gained control of the city.

The walls continued to stand as the city swapped hands during a series of civil wars in the 1300s, with invaders typically entering the city through bricked-up gates or doors left open by inside forces.

In total, of the thirty-five major sieges that took place before 1453, twenty-four failed. And the walls stood through all of them. They repelled invaders from the Ottoman Empire three separate times: in 1391, 1394, and 1422.

That all changed in 1453.

Imagine the shock and terror felt in the city when the residents heard the otherworldly roar of a half-ton cannonball flung from a cannon like no other striking their walls like the hammer of an angry god. Defenses that had lasted for more than a thousand years lasted a mere forty-seven days. This was not a mere military loss. It was essentially the end of the Byzantine Empire, which traced back to the early Middle Ages.

The power of disruption. One whose complex power remains to some degree a mystery to science even today. Fire drug. The devil's distillate. *Fuegos artificiales*. One whose power motivated Shakespeare to write, "And O you mortal engines, whose rude throat / Th'immortal Jove's dead clamours counterfeit."[1]

Gunpowder.

A Brief History of Gunpowder

Saltpeter, sulfur, and charcoal. Sulfur reacts with the heat from a spark of a flame. It ignites at around 261°C, generating additional heat, which ignites the charcoal. The saltpeter (potassium nitrate or sodium nitrate) shatters and releases its oxygen. The oxygen ignites more fuel, accelerating the burning, a process known as *deflagration*.* The charcoal burns, emitting thermal energy. A fuse, an enclosed container, a tube, a rocket . . . well, you know the rest.

Gunpowder transformed battle. It reshaped society. Sir Francis Bacon, whom we will meet again in chapter 3, said that it (along with the printing press and the compass) "changed the appearance and state of the whole World."[2] It created a line between before gunpowder and after gunpowder. But where did it come from?

Well, the truth is, no one knows exactly. We can see its beginnings through a—pardon the pun—haze of smoke.

Historians generally believe that gunpowder was discovered and developed in China. Early references came from exactly where you would expect it to come from.

Philosophy books.

The first reference to something that sounds like gunpowder is in the 142 book *The Book of the Kinship of Three*, which referenced how combining three powders would lead to sparks that "fly and dance violently."[3] Author Wei failed to mention what those three ingredients were, so we're not there yet.

Two centuries later, we see the outlines of a formula in Ge Hong's 318 *Book of the Master Who Embraces Simplicity*. Ge's formula had saltpeter, pine resin, and carbonate but didn't describe ratios or how to combine the ingredients. Near the end of the next century, prolific author Tao Hongjing called the violet flames that come when saltpeter is burned "a miraculous product of heaven and Earth" in *Informal Records of Eminent Physicians*, one of eighty books he wrote.[4]

Something is happening here, but what it is isn't exactly clear. You might picture long-robed philosophers stroking their beards, running

---

* My longtime Innosight colleague Andy Parker named me a world-class nonsense peddler (he used a different word, but this is a G-rated book!) because of my occasional use of big words. I admit to liking to use *conflagration* on occasion. I also admit to never having heard of the word *deflagration*.

experiments, seeing purple flames, and then returning to philosophizing. It isn't clear why people are experimenting, what they hope to learn, and what they do with that knowledge, other than to have more reasons to stroke their beards.*

What felt like pure pondering turns into more purposeful experiments done by alchemists. What could this mysterious substance saltpeter do? It certainly could be explosive. In his 652 book *Prescriptions Worth a Thousand Gold*, Sun Simiao described how to fill a pot with pulverized saltpeter and sulfur, place the pot in a pit, add three soap horns (whatever those are), add a handful of bark shaving from a tree, and add and then light three pounds of charcoal. That process led to gray smoke and a powerful flash. Now, Sun was known as the "King of Medicine."[5] And the quest to find ways to live longer, better lives—one of humankind's timeless quests—is where the story begins to accelerate.

Chinese practitioners had been working on elixirs meant to elongate life. In around 850, an alchemist whose name is lost to history blended saltpeter, sulfur, and charcoal in a pot. The book *Gunpowder* by Charles River Editors vividly describes what happened next:

> *In his sleepless daze, he inadvertently elbowed a lit torch and knocked it into the receptacle, which produced a thunderous explosion that reduced half of his workshop to rubble. Fortunately, the experimenter lived to tell the tale and promptly shared his findings with his peers. Alchemists worked unceasingly to tweak the measurements and eventually agreed upon the following ratios: 75 parts saltpeter, 15 parts charcoal, and 10 parts sulfur—thereby creating the first official gunpowder formula.*[6]

Around the same time, the wonderfully titled *Classified Essentials of the Mysterious Tao of the True Origin of Things* referenced elixirs

---

* No, I have no idea if there was any beard stroking going on, but it's what I see.

that had all the vital ingredients of gunpowder. The mixture was considered "a dangerous mistake" that when ignited "burns with a soft whomp and a burst of flame, leading behind a cloud of dense white smoke—a magician's effect."[7]

That soft whomp and burst of flame. For the next two hundred years, gunpowder was largely a tool for illuminators and magicians. A monk named Li Tian tossed a series of bamboo shoots with gunpowder into a fire and, voilà, firecrackers!

Leaders, as they are wont to do, began in parallel to experiment with potential military uses. The first "fire arrow" appeared in 904. In the next century, scientists began to experiment with bombs such as the "flying cloud thunderclap bomb." The big development in the twelfth century was gunpowder-powered projectiles, cylindrical bamboo barrels propped up by poles to fire explosives. The first of what we now call a cannon appeared at the Siege of De'an in 1132.[8] The Chinese kept working on new versions of the hand cannon. By 1412, they had innovated their way to the "long-range awe-inspiring cannon." Or, you might prefer another version, the "nine-arrow heart-piercing magic-poison thunderous fire eruptor."

Somewhere in that 500-year period separating the soft whomp of a failed experiment to create a life-extending elixir and the long-range awe-inspiring cannon, gunpowder made its way to Europe and transformed the battlefield.

Three Military Moments

It's not clear exactly how the idea spread to Europe. Perhaps it was from the Mongols, or from trade with the Arabs, who had either gained knowledge from China or developed their own gunpowder formulas independently.

There was even a parallel development of a mysterious substance called Greek fire, which might have been a historic precursor to napalm.

Legend had it Greek fire set anything it touched on fire and could not be put out by water. An account of a 941 battle described how "soldiers preferred to throw themselves into the ocean weighted down by their armor and weaponry rather than face the horrible liquid."[9]

The secrets to its mastery apparently died with its creators, as it now serves as a historical footnote in the development of modern warfare. Gunpowder, on the other hand, played an increasingly critical role. Three key military moments cemented its disruptive power.

The first moment happened in 1346 in Crécy, in France. It was during the Hundred Years' War between France and England. King Edward III, the thirty-four-year-old English monarch, had carried gunpowder with him to the battlefield. Jack Kelly describes the scene in his book *Gunpowder: Alchemy, Bombards & Pyrotechnics; The History of the Explosive That Changed the World*:

> *Astound. Astonish. Stun. Detonate. All of these words derive from a root meaning thunder. On firing, Edward's guns shot out great tongues of flame followed by rolling clouds of white smoke, an impressive and unique sight to the French knights and their allies. More astounding, more stunning was the powerful sound of the detonations. If, as one chronicler noted, it scared the horses, it most certainly startled the men as well. It was thunder brought to earth, sound hurled forth as a weapon.*[10]

Shock and awe, version one. Sieges shortened. Castles became less effective strategically. By 1415, Henry V had guns with names like "London," "Messenger," and "The King's Daughter."[11]*

Fast-forward to 1428. War still raged between France and England. King Charles faced a different kind of threat: a seventeen-year-old

---

* As the father of three boys and, at the time of this writing, a seventeen-year-old girl, I hear you, King Henry!

peasant named Jehanne, known as La Pucelle ("the Maiden"). Today we call her Joan of Arc. The young woman earned particular notoriety for her innovative use of artillery on the battlefield. Why was she able to see what more-experienced commanders could not? Because Joan was a peasant, she could mix easily with gunners. She could then pick up and incorporate new ideas. Joan also had the benefit of not having to unlearn old techniques before adopting new ones. As Kelly put it, "Military commanders versed in classical theories, for whom gunpowder weaponry was an awkward intrusion, struggled to incorporate the guns into their strategic thinking. Joan, lacking preconceptions, viewed artillery with fresh eyes and readily developed an intuition about its use."[12]

Joan, of course, was famously burned at the stake by the English, accused of being a heretic and sorceress. Questioning the status quo carries risks.

Our final military moment returns to Constantinople in 1453. It is the last days of the Byzantine Empire. At its peak, the empire covered much of the land bordering the Mediterranean Sea. But by the mid-fifteenth century, it had some land in modern Türkiye and Greece. A mysterious figure appeared. A gunner named Orban, sometimes known as Urban. Orban came from Hungary. Or perhaps what we now consider Germany. Regardless, historians agree that the skilled metallurgist and gunpowder technician offered his service to the Byzantines who ruled Constantinople. In return, the Byzantines offered a small salary and a meager supply of metal. Orban looked for the next bidder, Sultan Mehmed II of the Ottoman Empire. The empire had failed in its previous two efforts to take Constantinople. Mehmed specifically sought a gun to topple Constantinople's iconic walls.

"I can cast [a cannon] with the capacity of the stone you want," Orban reportedly told the sultan. "I have examined the walls of [Constantinople] in great detail. I can shatter to dust not only these walls with the stones from my gun, but the very walls of Babylon itself."[13]

Orban created a "giant bombard in two parts that screwed together. The rear section was a chamber thick enough to withstand the explosion of a massive quantity of powder; the barrel's bore was large enough to accept an enormous stone projectile."[14] That was just the beginning. His next gun, built in January 1453, had a twenty-six-foot barrel that could throw a stone ball weighing more than half a ton a mile (figure 1-1). To avoid panic among the locals, Orban gave them a day's notice before testing the gun. On April 12, Orban's weapons, sixty-nine regular cannons, dozens of catapults, and an 80,000-person force began its assault. Forty-seven days later, on May 29, after the Ottomans burned through fifty-five thousand pounds of gunpowder, the city fell.

"Neither man nor beast nor fowl was heard to cry out or utter a sound," one chronicler noted. "The city was desolate, lying dead, naked, soundless, having neither form nor beauty." Cardinal Basil Bessarion put it equally starkly: "The glory of the East, the refuge of all good things, has been captured."[15]

The gunpowder era would continue for the next four hundred years. Its end started in the 1860s, when Alfred Nobel patented the blasting cap. Nobel's cap enabled a chemical reaction of pure nitroglycerin so sudden that air molecules piled on top of each other, forming a shock wave that went faster than sound.* Nobel borrowed a word from Greek that meant *power*, and dynamite was born. Unlike the slow spread of gunpowder, the adoption of dynamite and its derivates by miners and construction crews happened much faster.

So, what took roughly a millennium to go from pondering philosophers wondering about mysterious materials to reasonably widespread application largely disappeared in decades.

---

* Yes, it's that Nobel. The story goes that when Alfred Nobel's brother died in 1888, one newspaper incorrectly filed Alfred's obituary and called him a "Merchant of Death." A great Marvel character, but not a moniker to which most humans would aspire. Alfred decided he would give the bulk of his fortune to establish prizes for notable contributions to literature, the sciences, and world peace.

Figure 1-1

The Dardanelles gun, cast based on Orban's gun in 1464

Source: © Royal Armourie, https://royalarmouries.org/collection/object/object-6177.

Gunpowder's Echo: The Challenge of Disruptive Change

The gunpowder story brings into sharper relief one of the key challenges facing leaders confronting disruptive change. Recall that the Byzantines had the chance to invest in Orban's cannon but chose not to. When you first see disruption, you don't feel the terror of a half-ton cannonball barreling down on you. Rather, your first instinct is often smug dismissal.

For example, on November 12, 2007, the cover of *Forbes* magazine featured a white man wearing a collared shirt and jacket and holding a mobile phone to his ear. The man has a hint of a smile, maybe you would even call it a smirk, on his face. Squint, and above the bar code and the $4.99 price tag for the issue, you see who this grinning man is: Olli-Pekka Kallasvuo, chief executive. And the glaring headline in all-capital letters tells you of which company: "NOKIA: ONE BILLION CUSTOMERS—CAN ANYONE CATCH THE CELL PHONE KING?"

History judges that question, asked two months after the launch of Apple's iPhone, as ridiculous. Yet in this vignette, I can see parallels to the fall of Constantinople and a key challenge of disruptive change. The moats, walls, and defenses that have served a company so well when playing the old game are utterly useless as the new game unfolds. The leaders of Nokia ruled over a seemingly impregnable fortress. The launch of the iPhone in mid-2007 and phones powered by Google's Android operating system a few months later didn't immediately impact its business. But continued innovation from Apple and the Android ecosystem ultimately relegated Nokia and other industry leaders such as Motorola and Research in Motion (the manufacturer of the popular BlackBerry device) to also-ran status.

I have sympathy for the Byzantines who underinvested in Orban's gun. At the time, there was no historical evidence about patterns of disruptive change, so it was legitimately challenging to fully understand the full disruptive potential of gunpowder.

I have less sympathy for Nokia. In 2014, INSEAD Professor Quy Huy interviewed Kallasvuo. The CEO answered Huy's initial question about what it was like to face the competitive dynamics of his time:

> *Boy. Yes. There was plenty of disruption and turmoil in my time as well. I think it basically relates to the fact that never before in business history has the competitive environment changed as much as it did in the converging industry of mobile communication, internet, computers at around 2006, '07, and '08. For everybody who was playing in any of these industries, these were really important, difficult, and also volatile times. And Nokia definitely got its fair share of the volatility and change.*[17]

Kallasvuo goes on to say that Nokia's strategy, which in essence involved clinging to its old model and making fitful investments in new software platforms, was right. The organization just couldn't execute. "It is sometimes difficult in a big successful organization to have the sense of urgency and hunger," he said.

I wasn't inside Nokia at the time, but the company's experience just so squarely fits the pattern Clay observed. The market leader discounts the potential of the disrupter, waits too long to allocate resources, and, when resources are allocated, fails to reimagine the business in the face of the change. And it's much harder to have sympathy for those mistakes. Nokia's struggles came after the mainstreaming of the idea of disruptive innovation.

You see, among the endorsements on the jacket of Clay's 2003 book *The Innovator's Solution* is one saying that Clay and his coauthor Michael Raynor "have done a superb job of creating a framework for helping to understand industry dynamics and for planning your own growth alternatives."

That comment was written in 2003 by Pekka Ala-Pietilä. From 1999 to 2005, Ala-Pietilä served as the president of Nokia.

Who Is the Hero of the Gunpowder Story?

A magician's effect, something likely discovered accidentally. Like all innovations, it is something unexpected, a break from the past. One that, at least initially, the world couldn't put to practical use.

Some people find things that are strange or unfamiliar to be scary. Disruptive innovators seek and savor the strange stuff. They question the status quo, looking for different and better ways to do things. "Most businesspeople, most people in any walk of life, tend to think that today is going to be pretty much like yesterday," historian Richard Tedlow notes. Disruptive innovators "think anew and act anew," orienting "toward a future beyond imagining to others."[18]

Disrupting the status quo requires looking at the world differently. A disruptive innovator sees what others don't. They do what others won't. Consider Kelly's description of the people dabbling with "the precious and poorly understood material":

> *The men who assembled its materials were the brothers of bakers and brewers: They devised their methods through intuition and knew that minor variations in procedures could significantly alter the outcome. The craft attracted adherents across Europe: alchemists, blacksmiths, enterprising peasants, men fascinated by the unknown or intrigued by the commercial potential,* **daredevils, visionaries, madmen**. *Some found in the profession not fortune but disfiguring burns and death—the grinding of gunpowder was ever a perilous craft.*[19]

One of the four questions I had in my head when I started this research is who actually does the work of disruptive innovation? Is it the lone genius? Or legions of faceless innovators who somehow coalesce around a disruptive idea?

The gunpowder story doesn't have a singular hero. Its development is sprawling, taking place across centuries. Alchemists, blacksmiths, enterprising peasants. We have seen military commanders, philosophers, alchemists, and gunners. There were scientists trying to tame the material. There were farmers and fighters experimenting with different uses. There were leaders allocating time and money and directing work. And that's for gunpowder itself. There are parallel stories, as there are in any disruptive development.

Consider saltpeter, one of gunpowder's essential ingredients. As mentioned, saltpeter is otherwise known as potassium nitrate or sodium nitrate. It is a white crystalline solid that tastes salty and bitter. You can find it naturally in cave walls or dark boulders. You can find it in pig pens or stables. You can extract it from the skin that forms on marshes when temperatures drop in the fall. Or you could produce it in a process that the Charles River *Gunpowder* book called "excretia." Here's a memorable description of that process:

> *Fresh manure was melded with ash and straw, then left to decompose, during which a crust of saltpeter crystals formed atop the manure-ash slab. The slab was then rinsed, draining off the crystals. Another method involved pouring an ash-tinged mixture of stale urine or cesspool fluids onto niter beds—rectangular blocks of rotted manure mingled with straw—repeatedly over the course of a few weeks. These damp beds were baked under the sunlight and would eventually be coated in saltpeter specks. Harvesting saltpeter was anything but glamorous, to say the least, and required individuals hardened to revoltingly pungent odors and fumes.*[20]

"Anything but glamorous" feels like an understatement. At some point, just about everything we do traces back to someone trying something and learning what works and what doesn't. They did something different that solved a problem or created value in some way. They *innovated* (see the sidebar "What Exactly Is Innovation?" for more about innovation). I often wonder about the person who first thought that a hard-shelled insect wandering in the water would make a delicious meal. Or came up with the idea of intentionally growing mold on cheese and eating it. Daredevils, visionaries, madmen, indeed.

And for ideas to spread, well, they have to spread. Gunpowder's development primarily came from trial-and-error experimentation. The more people could learn from experiments that had already been run, the more they could advance the practice of gunpowder.

As gunpowder became more widely available, so too did books about how to tame the incendiary device. In about 1280, Hasan al-Rammah wrote one about "treats of machines of fire to be used for amusement or for a useful purpose." One of the most famous ones was the indispensable how-to manual *Books of Fire for the Burning of Enemies* by Marcus Graecis (alternatively known as Marcus Graecus or Mark the Greek), various editions of which appeared in the early fourteenth century.*

Those manuals were handcrafted. The laborious process limited the spread of this knowledge. The slow spread meant that it wasn't easy to learn from the mistakes of others, so you had to make your own mistakes. And if a mistake was too bad . . . well, you didn't make it into the history books.

There was no choice. Ink, paper, tracing, movable type . . . pieces of what ultimately would be a different disruptive solution existed at the time, but their combination didn't. The printing press, which we'll describe in the next chapter, changed all that.

Vannoccio Biringuccio, born in Italy in 1480, created the first printed book dealing with the gunner's arts. The first edition of *De la pirotechnia* was published in 1540, a year after Biringuccio died and 100 years after Johannes Gutenberg had developed a working version of his printing press. Over the next 138 years, eight other editions would be published. Authors distilling and spreading knowledge helped hasten gunpowder's spread.

Gunpowder starts as a nameless, faceless story. As time progresses, specific names appear, but details, connections, remain hazy. The story has gaps of decades, sometimes centuries. There is a lot of inference, speculation, and guesswork. And who is the protagonist? Daredevils and visionaries, yes. But who specifically are these daredevils and visionaries? It is not clear. The words are plural, so we imagine there are many innovators, but we don't exactly know who they are or what motivated them.

---

* I think we've lost the art of evocative titles. *Books of Fire for the Burning of Enemies* would just pop on the shelf, don't you think?

What Exactly Is Innovation?

Austrian philosopher Ludwig Wittgenstein once noted that "the limits of my language mean the limits of my world." One key to understanding complex phenomena is having a common way to describe them. Despite its ubiquity in modern times, innovation often suffers from a lack of a common definition. That can lead organizations focusing on innovation to struggle with it, because everyone is, in essence, talking past each other. As the brothers Chip and Dan Heath note in their excellent book *Switch*, "What looks like resistance is often a lack of clarity."[16] Complex challenges require a common language to make sure that people can have productive discussions without talking past one another.

Having reviewed relevant academic and sociological literature and reflected on two decades in the field, I have landed on a simple, practical definition of innovation: *something different that creates value*.

That phrase breaks into two parts. "Something different" is intentionally vague, leaving space for both world-changing innovations (can someone please develop hypersonic planes?) and day-to-day innovations that make life just a little better. That broad definition is a reminder that innovation is not the job of the few; it is the responsibility of the many.

"Creates value" distinguishes innovation from precursors like creativity or invention. Those are critically important ingredients, no doubt, but until that spark turns into increased financial performance, more-satisfied employees, happier stakeholders, or some other form of value, you have not innovated.

That's it. Innovation is something different that creates value.

OK, back to your regularly scheduled program, already in progress . . .*

---

* I said that to my kids once, and they looked at me. I associate that phrase with a sporting event overrunning its scheduled block, leading to the network announcing that it was going to join the otherwise-scheduled programming in progress. Of course, today the concept of a scheduled program is increasingly a historical relic. Innovation marches on.

The stories ahead are different in one critical way in that there is a hero or, sometimes, several heroes. Yet they are also similar in that the hero never works alone. The hero has collaborators, coworkers, coconspirators. There are supporters. There are also doubters who can sometimes be dour or downright hostile. Disruption is a difficult pursuit but not necessarily a lonely one.

2

The First Information Revolution

How the Printing Press Changed Everything

In 1455, two years after the fall of Constantinople, another battle brewed. This one wasn't about who had the biggest cannon or the strongest walls. It was about a perhaps deadlier topic: money.

In one corner sat an inventor who could sometimes get distracted and didn't always quite deliver against his commitments. In the other corner sat the inventor's financial backer, who claimed the inventor had failed to fulfill the terms of his contract.

The court carefully considered both sides before deciding in favor of the investor. The inventor, who had just delivered something improbably difficult, something that sits in museums today and that seemed primed for runaway commercial success, had to cede his technology to the investor. The investor allowed the inventor to stay commercially active, but over the subsequent years, the inventor's name began to fade from memory, with the investor and his business partner credited with leading a technological revolution with world-changing impact.

A couple centuries later, the inventor, Johannes Gutenberg, recaptured popular attention. "The lead in typecases has done more to change the world than the lead in bullets," remarked eighteenth-century German physicist Georg Christoph Lichtenberg.[1]

The printing press transformed religion, arts, the sciences, and more. It enabled the disruption of the creation and dissemination of knowledge itself. It allowed people like Biringuccio to spread knowledge about what used to be crafts like gunpowder. It allowed Martin Luther to change the world of religion. And it allowed philosophers and scientists to radically rethink the very nature of knowledge, a topic explored in the next chapter.

The investor, Johann Fust, and his printing partner Peter Schöffer, have now faded from memory.

History is, as always, fickle.

A Brief History of the Printing Press

Fiction sometimes reflects reality. In 2005, Scott Brown and Anthony King wrote and composed *Gutenberg! The Musical!* The protagonist creates the technology overnight, fails to notice that his beautiful assistant Helvetica is in love with him, and fends off an evil monk named . . . Monk . . . who worships the devil and wants to destroy the printing press to control locals through inaccurate information.* The

---

* I never saw the musical. Its first run didn't officially make it to Broadway. A revival ran on Broadway from October 12, 2023, to January 28, 2024. Wikipedia tells me that Triumph the Insult Comic Dog was one of the notable viewers of the show. And now you know.

plot is obviously ludicrous, but there is ample space to take creative license with the Gutenberg story.

While historians universally agree with the basic story, there are significant and wide gaps in specific details. Some of those gaps span decades. The lack of documentation for much of Gutenberg's life reflects, ironically, the very challenge of documentation that Gutenberg disrupted.

The story starts in Mainz, Germany. Gutenberg was born somewhere between 1394 and 1404. The world picked 1400 as Gutenberg's birth year because, well, who doesn't like round numbers? John Man describes the scene in *The Gutenberg Revolution*:

> *In 1400 modern concepts of scientific and historical truth hardly existed—sources were as rare as desert flowers, to be found, if at all, only by a lifetime of travel. The only truth was that of the Church, which, like Big Brother, controlled the media, in the form of scribes (for the written word), priests (oral transmission) and artists, who served both. The Church had become rich beyond imagining, with the faults that wealth and privilege bring.*[2]

The craft of writing was about three thousand years old, and paper (which, like gunpowder, originated in China) was about a thousand years old. Knowledge capture and transfer was a slow, laborious process. The typical production of manuscripts involved skilled scribes painstakingly writing on parchment. Woodcut printing, which had emerged in China, Japan, and Korea several centuries before paper, was primarily used for playing cards, pictures of saints, or short pamphlets. A particularly popular genre for woodcuts was *Ars moriendi*, or "The Art of Dying." These woodcuts used vivid illustrations and short text to show "the struggles with death, the temptations of the Devil and redemption through Christ."[3]

The limits of scribes and wood carving meant that information spread glacially and inconsistently. Libraries were small, as was the number of people who could read. Storytelling remained one of the most critical ways in which information traveled. And the Catholic Church dominated Europe. It also dominated information, as the Church was perhaps the only organization that had the budget and desire to sponsor the painstaking, manual process of creating and copying written materials.

Gutenberg came from a relatively well-off family. His father served as "Companion of the Mint," a position that historians think might have exposed Gutenberg to hand-pressed coin making.

The basic process of casting a coin involved making a mold from a die that had been indented with a punch. The raised patterns of the punch, transferred to the die and imprinted on the mold, allowed coins to have their distinctive markings. Coin making was a true art, with a skilled coin maker able to carve a letter in steel that was up to sixty times sharper than the resolution of a modern laser printer.[4]

Why did a skilled coin maker live and work in Mainz, a relatively small city with only six thousand inhabitants? In the fifteenth century, every sizable region had its own form of currency, so each region had to have someone who knew how to press coins. With skilled craftspeople gathered in guilds, knowledge could spread, albeit slowly.

From Mirrors to "Arts and Adventures"

Little is known about Gutenberg's early years. Gutenberg had some inheritance, but it was limited, and a system of patronage that benefited the upper class in Mainz teetered in the early part of the century as the city suffered from rampant civil strife. Limited inheritance meant that Gutenberg had to work for a living. He apparently went to school. It seems he made friends and dabbled in a range of business interests. He may or may not have been kicked out of the city.

The First Information Revolution

As a young man, he was not an obvious candidate to change the world. And yet the hunger to upend, unseat, transform ... disrupt, appears. "He emerges," Man writes, "as that rarity: a man seized by an idea, obsessed by it, *imprinted* by it, who also has the technical skill, business acumen and sheer dogged, year-after-year grit to make it real."[5]

Making it real started in earnest in 1434, when Gutenberg moved to the outskirts of Strasbourg. At the time, the city had about twenty-five thousand inhabitants. It had vibrant bell-casting and paper manufacturing industries and robust trade links to France, Italy, Prague, and eastern areas of Germany. Historical fragments show that Gutenberg was moderately well-off. A tease of a 1436 note suggests that Gutenberg might have broken a promise to marry Ennelin zu der Iserin Thüre. When recordkeeping is slow and laborious, few documents are recorded, and historians remain busy with multiple mysteries forevermore.

In 1437, Gutenberg started tutoring Andreas Dritzehn in the craft of minting and goldsmithing. The two partnered to create a company to finance the development of a new, unnamed technology, which may have been a precursor to the movable-type printing press. In parallel, in 1438 Gutenberg entered into a contractual agreement with Strasbourg bailiff Hans Riffe von Lichtenau to produce mirrors of polished materials that pilgrims would use to capture holy rays during the 1439 Aachen Pilgrimage.

Yes, it then took more than a year of planning to produce mirrors made from a basic alloy of lead and tin. And yes, people believed a reflective object could absorb the essence of the Holy Spirit. Historians generally place the end of the Dark Ages at sometime between the tenth and fourteenth centuries, but you can see echoes of it in this vignette.

So, Gutenberg was working on mirrors. And as so often happens in these stories, life happens. The pilgrimage ended up being delayed because of an outbreak of bubonic plague. A century earlier, the black death had ravaged Europe, killing by some estimates about half the population, so it's easy to see how the new flare-up rightfully raised a lot of fears.

With the trinket plans delayed, Gutenberg changed course. In 1439, Gutenberg, Dritzehn (who died that year), and Riffe von Lichtenau signed another contract to work on Gutenberg's "arts and adventures," which, according to Professor Stephan Füssel, "was a guildsman's technical term for adept artisanal skills and bold mercantile ventures."[6]

Details were vague. The contract referenced a wooden press made by Conrad Saspach. There are records of a large purchase of lead. A goldsmith named Hans Dünne received 100 guilders, or about $50,000 in modern terms, for vague fabrication work related to printing.* Pieces of a world-changing idea were coming together.

The Magic of Gutenberg's Printing Press

The exact timeline of technological developments over the next decade is unclear. By the end of the 1440s, however, Gutenberg and his team had done something remarkable. They had a device that could turn ink, die-cast metal type arranged to form sentences, and paper into printed material. And could do it again. And again. And again.

That device hinged on Saspach's screw press. At its most basic level, using a press involves putting an object between two plates and pressing the plates together to transform the object. A wine press turns grapes into liquids. An olive oil press extracts juice from olives. The waffle press in our kitchen turns gooey batter into delicious, flaky waffles.** You get the idea. Gutenberg's press upended the standard practice of rubbing ink over a woodblock or another carving. The press provided consistent pressure, which led to faster and better results. It would take a skilled scribe a week to copy two dense pages. Once a

---

* That's a very rough approximation. Guilders don't appear on modern inflation-adjustment tools.

** Waffle makers are amazing and truly mistake-proof. My thirteen-year-old son, Harry, repeatedly requests chocolate chips with a side of waffles. Dad happily obliges.

page was set, a printer using a press could create a copy of it significantly more quickly.

According to Füssel, Gutenberg's printing press combined *eight* distinct innovations:

1. The concept of individually molded letters

2. A mix of roughly 83 percent lead, 9 percent tin, 6 percent antimony, and 1 percent copper and iron for characters

3. Instruments to cast the type

4. Specific fonts

5. Composing sticks to enable precise alignment of type on a page

6. The specific mix of lampblack and resin for ink

7. Leather balls to apply ink to type

8. The press itself with a "sliding carriage, pin holes for perfect alignment of type areas, frames to protect the non-printed surfaces, cover, platen, and bar"[7]

While that list sounds simple enough, explaining exactly how the printing press works is, in the words of Man, a bit like trying to explain exactly how to ride a bicycle. One key part of the first element (individually molded letters) was a handheld mold that enabled the rapid creation of letters struck into pieces of metal, which became known as a matrix. Here's Man's vivid description of a museum staffer demonstrating the mold at the Gutenberg Museum in Mainz.

> *His performance is like a magic show, complete with a puzzling gismo—the hand mould itself—and a retort of molten metal— mostly lead, with additions of tin (to increase the flow and the*

speed of cooling) and antimony (to harden the metal and thus ensure the sharpness of the letter). He doesn't let people too near the molten metal, not simply because of the dangers of its 327°C (621°F) [temperature]. . . .

. . . All you need to know is that the hand mould has two parts which slide together to grasp the matrix, with the imprinted letter facing upwards. With the matrix held firmly in place by a springy metal loop, the two parts leave a rectangular slot, at the bottom of which is the matrix with its imprinted letter. [The museum staffer] takes a ladle, scoops up molten metal, pours an egg cup's worth of it into the slot, lifts off the spring, slides the mould apart, and out falls a little silver rectangle, just over four centimeters long, already cool enough to hold. . . . This is a piece of type, or "sort" with its letter standing proud at the top. The whole operation takes less than a minute. . . .

. . . Here was a device, easily used, which could produce all of the amount of type needed for a book from a single set of punches. Later, expert typefounders could make four "sorts" a minute—several hundred (the alleged but unsourced record stands at 3,000) in a day.[8]

Making the printing press function required lots of tinkering. Type needed to be stored. Gutenberg engraved and cast 290 different characters: 47 uppercase letters, 63 lowercase letters, 92 abbreviations, 83 combined characters (such as *st*), and 5 forms of punctuation.[9] The right paper needed to be obtained. If the paper was too moist, it would dissolve. If it wasn't moist enough, it couldn't hold ink. Methodical trial-and-error taught the printers to dampen and put alternate sheets under pressure for a few hours. Ink needed to be stored. The process needed quality control. Printed material needed to be commissioned and distributed.

A decade of hard work gave Gutenberg what we today would call a proof of concept. That was an amazing accomplishment, no doubt. Even more remarkable acts were to come.

Fust Funds a Bet on the Bible

A proof of concept demonstrates that an idea works. Gutenberg and his partners weren't pursuing arts and adventures for sport. They clearly were seeking to create a commercial venture. This goal requires having an honest-to-goodness paying customer and investing to turn a laboratory experiment into a real operation.

The first customer Gutenberg thought of, naturally, was the Church. The organization had deep pockets and already produced and disseminated information. What problem did the Church have that the printing press could help solve?

Gutenberg started with a connection he made back when he lived in Mainz: Nicholas of Cusa. An influential figure in the Church, Nicholas became a cardinal in 1447. He was an interesting historical figure. His thinking was rooted in his *De docta ignorantia*, or "On learned ignorance," which held that the purpose of learning is to learn how little you know when seeking God. Nicholas drew energy from seeming paradoxes and had what historians called an obsession with making opposites coincide.* Perhaps the concept of the printing press appealed to him because it was an apparent contradiction, as it was an inventive way to standardize materials. Perhaps even more importantly, Nicholas had a problem to be solved.

At the time the Church was looking to standardize missals, or the basic service that was used during mass. The printing press was ideally suited for standardization. Scribes were slow, errors could creep in, and any two missals could look just a little different. The printing press allowed for perfect replication.

---

* I love paradoxes—in 2021, I wrote a master's thesis titled "The Paradox of Paradoxes"—and finding someone interested in paradoxes is interesting to me and is perhaps interesting to some of you at least. If you too love paradoxes, contact me at scott.d.anthony@tuck.dartmouth.edu, and let's chat.

What to perfectly replicate, however? The standard that existed in Rome? Or the standard that existed in Nicholas's church? It turns out there were material differences between the two. One imagines the knight in the movie *Indiana Jones and the Last Crusade* telling Harrison Ford, "You must choose, but choose wisely," as (in the movie at least) one option promised eternal life and the other immediate death. Gutenberg chose neither. Instead, he executed what modern startups would call a pivot. Keep the Church as a customer but shift to printing a Bible.

A missal was relatively short and would have been a reasonably trivial challenge to print. A Bible, however, was a tough challenge, maybe the toughest conceivable challenge for Gutenberg. He would have to produce something that was as beautiful as scribal Bibles. Maybe even more beautiful. It would have to be as accurate as scribal Bibles. Maybe even more accurate. And he would have to produce something that would be close to thirteen hundred pages.

The Church, however, always needed new Bibles. It would take a scribe about three years to create a single Bible. Gutenberg's press created the tantalizing possibility of creating and spreading scripture at unprecedented pace and scale. And there was no debate about which Bible to print.

Gutenberg couldn't just snap his hands and produce a Bible. He needed to hire labor, build another workshop, and get raw materials.

There is no documented contract between the Church and Gutenberg. In fact, it's not even clear that the Church agreed to buy a certain number of Bibles or if Gutenberg took it on faith, pardon the pun, that if he created Bibles, he could sell Bibles. Regardless, it's safe to assume that if there was a down payment, it wasn't enough to complete the project.

Although Gutenberg and his team had supported themselves while spending a decade developing the printing press, commercializing it would clearly require more money than they had. The bet on the Bible required financing.

There was no venture capital industry to fund a speculative startup, so, in 1449, Gutenberg first took a 150-guilder loan (about $75,000 in today's terms) from a relative and then took a larger loan of 800 guilders ($400,000) from Johann Fust, a local goldsmith and merchant. Fust and Gutenberg apparently didn't share relatives or previous business dealings or mix in the same crowds, but each found something in the partnership . . . for a while. The loan was specifically to build tools for the "Work of the Books." It carried a standard interest rate of 6 percent, with failure to comply leading to Fust taking over the tools his funding supported.[10]

It ultimately took four years to complete the Bibles. Along the way, other commercial opportunities emerged. For example, the Church would regularly run campaigns where it sold indulgences, which essentially absolved the purchaser from sins (and funded important Church priorities).

Gutenberg printed leaflets supporting a 1452 campaign designed to fund Catholic crusades in the Middle East, potentially for his old friend Nicholas of Cusa, who granted a local authority the right to sell two thousand indulgences in Frankfurt. When gunpowder hastened the fall of Constantinople the next year, Gutenberg printed a calendar that provided warnings about Eastern invaders.

It isn't clear why Gutenberg said yes to these smaller opportunities. Perhaps he wanted a bit of extra cash. Perhaps he wanted lower-stakes ways to sharpen the production process before he jumped into his big project. Perhaps he was just easily distracted. Whatever the reason, he found himself behind in his loan payments to Fust and needing more money to build a second printing workshop. Fust was skeptical but agreed to a second tranche of funding of 800 guilders, this time as a capital investment in the business instead of as an interest-bearing loan.

Successes and Failures

The year was 1454. It was primed to be Gutenberg's moment. The arts and the adventures had turned into a functioning printing press. He had demonstrated its usefulness with leaflets and calendars. A team of thirty had worked across three presses over the several years to create two-volume, three-million-character Bibles.

And those Bibles.

In the end, Gutenberg created about 150 of them. About a quarter of the total were printed on vellum (12 vellum copies survive), a number that would have required about five thousand calfskins to be shaved, softened, treated, stretched, dried, and scraped smooth enough to take type. Each of those five thousand calfskins would take about a month to prepare. The rest of the Bibles were printed on paper imported from Italy.

Six typesetters left their indelible mark on the pages of the Bible. Each page followed scribal traditions with two text columns and broad decorations (which were done by hand). Each page did something no scribe could do: the text on the right-hand margin was justified (figure 2-1). At first, Gutenberg had forty lines per column. He paused printing in the middle and experimented with forty-one, and then forty-two, lines per column (the more lines, the fewer pages required). Today the Bible is known as the forty-two-line Bible, or B42 for short.

Enea Silvio Piccolomini, who went on to become Pope Pius II, wrote to a Church colleague about seeing the Bibles in Frankfurt in 1454, noting how they were "executed in a very neat and legible script" that could be read "without the use of glasses." The bet on the Bible looked like it was a success, as Piccolomini told his colleague that he would like to buy a copy but that "there was already an ample sufficiency of buyers lined up . . . even before the volumes came off the press."[11]

Fame and fortune for Gutenberg? Fame that continues to this day, yes. Fortune, no.

In the middle of 1455, with the Bibles hot off the press, Fust sued him, saying that the entire process had taken too long and cost too much. Fust won. He had already built his own, competing workshop and now took over one of the workshops Gutenberg had built. Critically, the acquisition also allowed him to take over employment of one

Figure 2-1

The Gutenberg Bible

Source: NYC Wanderer (Kevin Eng), "The Gutenberg Bible," in *Mid-Manhattan Library* (album), May 28, 2009, https://www.flickr.com/photos/starfire2k/3631902258/in/photolist-6wWs1s-7Q2EEA.

of Gutenberg's master apprentices, Peter Schöffer, who just so happened to be married to Fust's daughter Christine. Fust allowed Gutenberg to keep a single production facility and a share of profits from the Bible. "In the end," Man wrote, "as bastards go, Fust was less than a complete and utter one."[12]

A few years later, in 1462, Mainz was the site of a skirmish over competing religious interests. Fust and Gutenberg—now commercial rivals—both jumped in. The two competitors avoided picking sides in the propaganda war, eagerly picking up contracts with both sides. Participating in that propaganda battle was, in essence, Gutenberg's last commercial gasp, with the Fust and Schöffer partnership thriving and expanding.

Gutenberg died in 1468. An investigation by a German professor in the 1640s began to shift attention from Fust and Schöffer to

Gutenberg. The 1741 publication of "A Well-Deserved Rehabilitation of Johann Guttenbergs [sic], as Amply Attested in Surviving Documents" by historian Johann David Köhler of the University of Göttingen cemented the narrative that Gutenberg was the hero of the printing press story. A celebration of what might have been his five hundredth birthday in 1900 gathered crowds from around the world. In 1999, a panel of experts proclaimed Gutenberg the man of the *millennium*: "If not for Gutenberg, Columbus [ranked no. 2] might never have set sail, Shakespeare's genius [no. 5] could have died with him, and Martin Luther's *Ninety-Five Theses* [no. 3] would have hung on that door unheeded.... The printing press ... helped spread truth, beauty, and, yes, heresy throughout the world."[13]

The Magic of Intersections

The printing press story has its moments. A venture foiled by the plague. A paradox-pondering customer. A last-minute breakup between financier and disrupter. How to properly apportion credit for the printing press's impact will remain a mystery forevermore. Its impact, however, is undeniable.

Remember how Sir Francis Bacon said the printing press, gunpowder, and the compass were three inventions where there was a clear before and after? Before, it took a month or two to create a copy of a book. After, hundreds of copies could be created in a week. Before, the annual production of Europe's book industry, as it were, could fit in a single wagon. After (by 1500), there were tens of thousands of titles and millions of volumes.

The printing press, as Man notes, changed everything: "Hardly an aspect of life remained untouched. If rulers could bind their subjects better, with taxes and standardized laws, subjects now had a lever with which to organize revolts. Scholars could compare findings, stand on

The First Information Revolution

each other's shoulders and make better and faster sense of the universe. Gutenberg's invention made the soil from which sprang modern history, science, popular literature, the emergence of the nation-state, so much of everything by which we define modernity."[14]

There's a question hiding in the background of the story. Why was it *Gutenberg* in *Germany*?*

The components of the idea largely existed before 1440. Ink wasn't a new idea, though Gutenberg did develop a specific mix of lampblack and resin. Materials that could capture ink wasn't a new idea. Ancient Egyptians had used papyrus to capture writing, and the Chinese pioneered papermaking during the Eastern Han dynasty in about 100 CE. Papermaking spread to Korea and Japan a few hundred year later and to the Islamic world in the eighth century. It spread to Spain in the twelfth century. Movable type wasn't a new idea; a Chinese blacksmith named Pi Sheng had conceived of the idea four hundred years before Gutenberg did his work. Using technology to speed the production of printed materials wasn't a new idea. The idea of stamping on paper emerged around the fifth century.

No, the pieces of the printing press weren't new. The *combination* of the screw press, movable type, ink, and paper into a device that could speed the production of printed material, however—that was something the world had never seen before.

With many of the initial ingredients bubbling around in Asia, it's natural to wonder why a European ultimately developed the printing press. At least one explanation is that the letters in European languages are much simpler than the characters in Asian languages. Perhaps it was a strong vocal tradition in the East, which led to less writing. Or maybe it was just plain old-fashioned luck.

---

* I mentally filled in "with the candlestick" at the end of that sentence. If you know, you know.

My view? Gutenberg's breakthrough illustrates one of the most persistent findings in studies of innovation, a key piece of the pattern of disruptive innovation: magic happens at intersections, when different mindsets, disciplines, backgrounds, and ideas collide.

Great innovators plant themselves at the intersections where collisions that spur magic happen. They remember the aphorism that none of us is as smart as all of us. The gunpowder story featured a hint of the magic of intersections: Joan of Arc's ability to mingle with gunners gave her insight into new ways to fight battles.

Intersections are all over the printing press story. Gutenberg's father worked on coin making. Strasbourg had a flourishing bellfounding industry, where skilled experts would carve blessings or manufacturing dates into the bell body. That location provided an ample stock of "technical experiences amassed by related crafts."[15] And the region of Germany in which Gutenberg lived and worked was replete with screw presses, used to make wine. As Steven Johnson notes in *Where Good Ideas Come From*: "By the mid-1400s, the Rhineland region of Germany, which historically had been hostile to viticulture for climate reasons, was now festooned with vine trellises. Fueled by the increased efficiency of the screw press, German vineyards reached their peak in 1500, covering roughly four times as much land as they do in their current incarnation. It was hard work producing drinkable wine in a region that far north, but the mechanical efficiency of the screw press made it financially irresistible."[16] (Side note: *festooned* is a beautiful word, isn't it?)

We see in Gutenberg's history a constant search for novelty. He left Mainz in search of new ideas. He worked with people with different backgrounds and skills. He shifted from trinkets to transformational technology. He ditched straightforward missals in favor of the complexity of the Bible. He kept pushing himself to the intersections. And he made magic happen.

The Printing Press's Echo: Disruption's Shadow

The printing press accelerated the spread of knowledge. That's generally a good thing. But it had downsides as well. It also accelerated the spread of what you could reasonably politely call propaganda and what you could less politely call manipulative, destructive lies. Also, the spread of knowledge had unanticipated knock-on effects.

The Catholic Church was thrilled with the power of the press when the innovation helped raise cash for the Church's crusades and get more Bibles into the hands of more people. Perhaps it was less thrilled when the press provided propaganda for both sides in Mainz's 1462 civil war. And, of course, many Church leaders likely cursed the technology the Church had helped create when rapid printing hastened Martin Luther's ability to spread the ninety-five points of disagreement he had with the Church, sparking the Reformation that splintered the Catholic Church's influence. A *third* of the books printed in Germany from 1518 and 1525 were from Luther.

"The noises that accompanied Luther's message of doom were probably not hammer-blows; they were the squeaks and bangs of busy printing presses," Man writes.[17] And the story replicated across England as the religious and civil landscape was transformed.

Perhaps the Catholic Church ultimately rued its decision to help spur an innovation that spurred the decline of its dominance.

Disruption casts a shadow. The printing press was a boon to some—scientists, revolutionaries, entrepreneurs who built businesses around it—and a curse to others: scribes, cardinals, and anyone else who profited from ignorance.

The participation of the Church in a technology that ultimately hurt it reflects a common theme in more-modern disruptive changes. There's often a perception that the fundamental problem incumbents have when facing disruptive change is myopia. That is, they simply don't see the change coming until it is too late. I have never once met a market leader that didn't see the disruption coming. In fact, in many examples, the incumbent, like the Catholic Church, plays a critical role in its development.

Consider Eastman Kodak. In the 1980s, Kodak was one of the world's powerhouse companies. Its rise to prominence began when it launched its affordable Brownie camera in 1900. In the decades that followed, Kodak established a dominant position in the lucrative film business, with its "you press the button, we do the rest" slogan demonstrating its commitment to making photography accessible to the masses. On January 19, 2012, Kodak declared bankruptcy.

The lazy story is that Kodak missed the shift to digital imaging. But the true story is more nuanced. The company created a working prototype of a digital camera in 1975. The engineer behind that project, Steve Sasson, offered a memorable one-liner to the *New York Times* in 2008 when he said management's reaction to his prototype was "That's cute—but don't tell anyone about it."[18] A cute line, but not an accurate one. Kodak invested heavily in digital imaging—billions of dollars—to carve out a strong position in the digital camera space with its EasyShare line of products.

What ultimately got Kodak was the merging of cameras and mobile phones. People didn't stop taking pictures, of course; they just shared them on social media instead of printing physical copies. And here's the thing. Kodak wasn't unaware of this shift, either. The company started the aughts with a bold bet—buying photo-sharing site Ofoto in May 2001.

Imagine a world where Kodak executives say, "What's our tagline again? Share memories, share life? Let's make it as easy as possible for people to share photos on Ofoto. Heck, why don't we let them share updates and links to news stories as well?" Mark Zuckerberg did just that when he created Facebook. Three years later. Imagine a world where early social networks were run by large corporations rather than freewheeling startups. That world might be better. It might be worse. It would certainly be different.

In this world, of course, that didn't happen. Kodak stuck its brand on the Ofoto website and then made it harder, not easier, for people to share pictures. Rather than try to reinvent and reimagine itself in the face of disruptive change, Kodak tried to get more people to print more pictures.

Disruption is hard. It threatens power dynamics inside organizations. Modern research by behavioral psychologists shows that people prefer *avoiding* a loss to *receiving* equivalent gains. It's no surprise then that disruption places a market leader in a real dilemma. Even if a leader rationally knows that investing in disruption is a smart long-term decision, the short-term pain of a transition from one state to another can lead to natural hesitation.

Disruption casts a shadow.

We can see this shadow through the lens of history in the story of the printing press. It was so obvious in the century after Gutenberg's disruption that it led to a dramatic decree by the king of England. Let's flip the calendars to the 1500s and see how.

3

Bacon (Not the Food) and Boyle

Laying the Foundations of Modern Science

It was 1548. King Edward VI was in the second year of what would turn out to be a seven-year reign. Like all monarchs, he had the ability to, among other cool things, issue proclamations. The ten-year-old king likely leaned on his advisers to issue an interesting one. Almost a century after the Bible rolled off the press, King Edward VI issued "A Proclamation Against Those That Doeth Innouate" (figure 3-1).*

Gutenberg's printing press allowed viewpoints to spread. Good for those who wanted to share knowledge. Bad for those who depended on knowledge being contained. Innovation is something different that

---

* Yes, I know that's not how we spell *innovation* today. Just like a quote from a British source would have grey skies and multi-coloured jackets and aluminium foil, we're going old-school in this chapter.

Figure 3-1

King Edward VI's 1548 "Proclamation Against Those That Doeth Innouate"

Source: Early English Books Online, Copyright © 2019 ProQuest LLC. Images reproduced by courtesy of Society of Antiquaries Library, London, https://www.proquest.com/books/proclamation-against-those-that-doeth-innouate/docview/2264202282/se-2 (accessed January 31, 2025).

---

creates value. Doing something different involves being curious and questioning the status quo. Questioning the status quo means questioning how things work, why they work the way they do, who makes decisions, why they have the right to make decisions, and so on. As viewpoints spread, it is not surprising that people who weren't used to being questioned didn't like being questioned so much.

Sociologist Benoît Godin dedicated his academic career to studying the history of the idea of innovation. In *Innovation Contested*, he noted that during Edward VI's time, "the innovator was a deviant, or rather, a defiant. He takes liberties in thinking and action contrary to what he has been educated for and contrary to the established order and orthodoxy."[1]

For centuries, Godin's research showed that innovation was frowned on. This attitude began to change in the seventeenth century. It was not a disruptive product or service that began to drive this change. It was a disruptive idea: the scientific revolution.

A Brief History of the Scientific Revolution

It's obviously overstated to say that knowledge didn't advance during what gets dubbed the Dark Ages. The stories of gunpowder and the printing press showed that it certainly did. Indeed, the spread of schools, the increase in literacy, and the formation of universities helped spread classical scholarship. Islamic scientists played a notable role in advancing the understanding of astronomy.

This research helped inform Nicolaus Copernicus's assertion that the Earth revolved around the Sun instead of the other way around. But discovery was the exception. The dominant view was that classic Greek philosophers covered what was needed to be known about science, and religious texts covered what was needed to be known about the world.

Indeed, Giordano Bruno built on Copernicus's work and argued that there could be other planets that supported life and that the universe was infinite, without a center. His reward? He was arrested and imprisoned while on trial for seven years. On February 8, 1600, he was sentenced to death. His alleged response was "Perhaps your fear in passing judgment on me is greater than mine in receiving it." Nine days later, he was gagged and burned alive at the stake.[2]

And people in modern times complain about ghosting and cancel culture. It might not always feel like it, but we've progressed.

The king's proclamation is written in the language of its time, but three parts of it stand out:

> *It tendeth bothe to confusion and disordre, and also to the high displeasure of almightie God, who loueth nothyng so muche as ordre and obedience.*

This point is consistent with teaching from the ancient Greek philosopher Aristotle—whose teachings remained foundational in 1548—that there is a natural order to things. Innovation, the proclamation reads, drives confusion and disorder, which displeases an order-loving God.

> *Wherefore his Maiestie straightly chargeth and commaundeth, that no maner persone, of what estate, ordre, or degree so euer he be, of his priuate mynde, will or phantasie, do omitte, leaue doune, chaunge, alter or innouate any ordre, Rite or Ceremonie, commonly vsed and frequented in the Churche of Englande.*

Don't mess with any Church of England ceremony. Don't even *think* about messing with any Church of England ceremony. You mess with a ceremony, you mess with the king.

> *Whosoeuer shall offende, contrary to this Proclamacion, shall incurre his highnes indignacion, and suffre imprisonment, and other greuous punishementes, at his maiesties wil and pleasure.*

"You come at the king, you best not miss," drug-dealing kingpin Omar Little proclaims in *The Wire*.* And if you tick off the king by innovating,

---

* *The Wire* is a tremendous TV show. It benefits from serial viewing because of the intricacies of its plot. I might be influenced by growing up a die-hard Baltimore Orioles fan in Maryland (the show is set in Baltimore), but to me, it is about as good as television gets.

you run the risk of being thrown in jail, beheaded, burned at the stake, or whatever the king feels like, basically.

Godin notes that, after the king's proclamation, "Trials and punishments followed. In the following century, documents by the hundreds made use of innovation to discuss the reformer as heretic, according to counter-reformers, using the word explicitly. Over a hundred of those documents made use of innovation in their titles, a way to emphasize a polemical idea and get a hearing."[3]

Now, the king's focus was on innovating religious ceremonies. But still, the proclamation captures a general mood. Changing the perception of innovation required, well, disruption.

Sir Francis Bacon

What is now commonly called the scientific revolution is, like any big shift in history, clearer in hindsight than when it was happening. It wasn't even known as the scientific revolution until French philosopher Alexandre Koyré used that phrase in 1939. There's no single moment, no single person, that marks before and after. Centuries later students still learn about individuals that played a part in the revolution, such as Galileo Galilei, Sir Isaac Newton, Robert Hooke, and Johannes Kepler.

This chapter focuses on Sir Francis Bacon and Robert Boyle because their work in combination provides both the theoretical basis of transforming knowledge development and a practical approach that would become the underpinning of scientific inquiry in the centuries ahead. Their disruptive idea was to question assumptions, rigorously experiment, and document and share what they learned. That idea led to innovation's radical shift from an epithet to a desired trait.

Here's a short personal digression about Bacon before getting into substance. He was married when he was in his mid-forties. His bride, Alice Barnham, was twelve days shy of her fourteenth birthday. It was a different time. A few months before Bacon died, he wrote Alice out of his will, noting, "Whatsoever I have given, granted, confirmed, or appointed to my wife, in the former part of this my will, I do now, for just and great causes, utterly revoke and make void, and leave her to her right only."[4] Ouch. After he died, it took her *eleven* days to remarry. Again, it was a different time.

Bacon had a remarkable career. Born in 1561, he was knighted in 1603 and became the solicitor general in 1607, the attorney general in 1613, and the lord high chancellor in 1618 (the same year he became a baron) before becoming the Viscount St. Alban in 1621. He served as a counselor to both Queen Elizabeth I and King James I.

And he thought. And he wrote. And he thought about what he wrote. And *how* he wrote.* One biographer notes a change in his handwriting sometime between 1600 and 1610 "from the hurried Saxon hand full of large sweeping curves and with letters imperfectly formed and connected, which he wrote in Elizabeth's time, to a small, neat, light and compact one, formed more upon the Italian model which was then coming into fashion."[5]

Bacon entered Trinity College at the University of Cambridge at the age of twelve, brimming with ideas about advancing knowledge. It took time for those ideas to take hold.

In 1604, he began writing *The Advancement of Learning*. The book argued that it would be a "disgrace" if, while humans were discovering the land, seas, and stars, "the bounds of the intellectual globe should be restricted to what is known to the ancients."[6] The book laid the foundation for advancing knowledge through reason instead of divine revelation. It was set to be published in October 1605. But as often is the case with best-laid plans, things happen. In this case, history happened. The Gunpowder Plot, a plan by Guy Fawkes and a group of other conspirators to blow up Parliament when the House of Lords

---

* Here's a fun fact. The lowest grade I ever got was in handwriting (a C). Sometimes I can't even read my own writing when I am marking up a draft. I actually wrote something different for this footnote but couldn't decode it. I'm serious.

and House of Commons gathered before King James I on November 5, 1605, dominated the public discourse, and Bacon's book found only a modest audience.

It took another fifteen years before Bacon published his disruptive treatise: *Novum Organum* (New organon, or instrument for acquiring knowledge). The treatise, part of a broader collection titled *Instauration Magna* (Great instauration) references Aristotle's classic work on logic. The cover page has an evocative image of a ship, which represents learning, sailing beyond the Pillars of Hercules.[7] The pillars were historically a landmark showing the limits of knowledge. A quote below the engraving translates to "many shall pass to and fro, and science shall be increased."[8]

Bacon's purpose was clear enough:

> *It is not possible to run a course aright when the goal itself has not been rightly placed. Now the true and lawful goal of the sciences is none other than this: that human life be endowed with new discoveries and powers. . . . Men have been kept back from progress in the sciences by reverence for antiquity, by the authority of men accounted great in philosophy, and then by general consent. . . .*
>
> *But by far the greatest obstacle to the progress of science and to the undertaking of new tasks and provinces therein, is found in this—that men despair and think things impossible . . . and therefore it is fit that I publish and set forth those conjectures of mine which make hope reasonable.*[9]

In *The Scientific Revolution*, an engaging overview of the era, Steven Shapin describes the transformational impact of Bacon's work: "The traditional expression of the limits of knowledge, *ne plus ultra*—'no farther'—was definitely replaced with the modern *plus ultra*—'farther yet.'"[10]

This change mirrored a critical event that took place more than a century earlier. In 1492, Queen Isabella of Spain agreed to back Christopher Columbus's journey to India. Columbus, of course, never made it to India, instead landing in what is now known as the West Indies. Further voyages by Columbus and other explorers ushered in the Age of Discovery and the race to colonize the "New World." And in the early 1500s, Spain replaced *ne plus ultra* on its coat of arms with . . . *plus ultra*.

A new era was emerging, one that valued discovery and questioning. "The renovation of natural knowledge followed the enlargement of the natural world yet to be known," Shapin writes.* "Practitioners of a mind to do so could use the newly discovered entities and phenomena to radically unsettle existing philosophical schemes."[11]

What does Shapin's observation mean? It starts with a belief that the world is understandable and that the path to understanding comes from disciplined experimentation. That didn't mean casting aside religious beliefs. Rather, many leading thinkers of the day believed that understanding the way the world worked revealed the beauty and majesty of God.

That view had held since Plato noted that "the world was God's epistle written to mankind" and that "it was written in mathematical letters."[12] Bacon echoed this view, saying that the quest for understanding is "after the word of God at once the surest medicine against superstition, and the most approved nourishment for faith, and therefore she is rightly given to religious as her most faithful handmaid."[13]

Nonetheless, questioning a belief system that largely guided much of the world since ancient times was a radical step. Bacon viewed traditional philosophies as so flawed that the world had "but one course left . . . to try the whole thing anew upon a better plan, and to commence a total reconstruction of sciences, arts, and all human knowledge, raised upon the proper foundations."[14]

---

* I bought a used copy of Shapin's book from Amazon. Whoever read it before me wrote (in very neat handwriting) "this is amazing" next to this paragraph. My contribution was (in my typically poor handwriting) "this is perfect." I didn't expect two footnotes on handwriting in this chapter, let alone this book. Innovation is full of surprises!

Bacon believed that it was important to understand cause and effect, in essence, the *why* of an observation. For example, a popular idea during his time was that you could heal someone injured in battle by treating the weapon that injured them with material from the injury. Bacon doubted the so-called weapon salve, because there was no *why* aside from pure mysticism.

Bacon suggested that people not bow to traditional views but seek to draw conclusions from their own personal experience. "Just as if some kingdom or state were to direct its counsels and affairs not by letters and reports from ambassadors and trustworthy messengers, but by the gossip of the streets; such exactly is the system of management introduced into philosophy with relation to experience," he wrote. "Nothing duly investigated, nothing verified, nothing counted, weighed, or measured, is to be found in natural history and what in observation is loose and vague, is in information deceptive and treacherous."[15]

He suggested that a good starting point to advance knowledge was a full register of all effects that could be observed in nature. A philosopher could then augment experience and natural effects with artificial experiments that could produce things hard to observe naturally. Experiments needed to be done carefully because the natural senses could deceive the experimenter. Experiments should have eyewitnesses. "Away with antiquities, and citations, or testimonies of authors," he said. Set aside "all superstitious stories" and "old wives' fables."[16]

Today we call this approach inductive reasoning. Make an observation. Determine patterns. Create a theory. This controlled experimentation was a large departure from the Greek philosophers because it disrupted the nature of an object, so in the words of one historian, experimentation "violates, rather than reveals, the nature of things."[17]

Bacon wouldn't live to see the full impact of his ideas. In 1621, two former business associations accused him of bribery and corruption.

Bacon, in frail health, did not deny the charges, and the House of Lords fined him £40,000 (about £10 million or $12 million today) and imprisoned him in the Tower of London. The king pardoned Bacon in 1623. Beyond the collection of essays mentioned here (and which also included several treatises on gardens; Bacon loved his gardens), he wrote *New Atlantis*, a utopian book that envisioned a modern university. Bacon died in 1626.

A remarkable man with a remarkable mind.* But for novel ideas to advance, they needed people to practice them, to show their usefulness.

Robert Boyle's Air Pump

One common element in many of the great startup successes of the past decades is the power of partnership. Steve Jobs and Steve Wozniak at Apple. Bill Gates and Paul Allen (and then Steve Ballmer) at Microsoft. Mark Zuckerberg and Sheryl Sandberg at Meta. Batman needs Robin. Simon needs Garfunkel.** There's power in pairs. Bacon on his own is a mere opining philosopher. For Bacon to have impact, he needed someone to put his ideas into action. After all, one of Bacon's biographers described his landmark book as "a work of immense erudition that can only appeal to, and be understood by, those who have already made a profound study of logic."[18] King James also noted the complexity of Bacon's work, saying it was "like the peace of God, that passeth all understanding."[19]

Enter Robert Boyle. In the late 1650s, Boyle's assistant Robert Hooke (who would go on to develop a microscope and coin the term *cell*)

---

* Some believe that Bacon wrote a number of plays for which William Shakespeare is given historical credit. I'll leave that theory for the real historians!

** Well . . . whether Simon needed Garfunkel can certainly be debated. Simon's solo album *Graceland* was one of the soundtracks of my youth. When I hear it, I am immediately riding in my father's old Saab, affectionately known as the "bucket of bolts."

developed an air pump.* Boyle was a leading member of an informal group of thinkers that formed the so-called Invisible College. In November 1660, the invisible became visible as Boyle and eleven like-minded philosophers created the Royal Society, dedicated to the advancement of knowledge. In 1662, 114 years after the king's proclamation, King Charles II signed a royal charter officially creating the Royal Society of London.

The presence of Boyle, a wealthy, well-connected Anglo-Irish aristocrat, helped make the pursuit of natural knowledge a respected one.

Back to the air pump. The device essentially pumped air out of a container, allowing an experimenter to simulate a vacuum. A reasonably modest device but a critical bridge between Bacon's philosophizing and the real world. Steven Shapin says the device was "emblematic of what it was to do experimental natural philosophy" and calls it "the Scientific Revolution's greatest fact-making machine."[20]

The idea was to use the pump to simulate what you would see at the top of the atmosphere. The pump fit Bacon's view that a human-made device could reasonably approximate nature. Shapin does a wonderful job of bringing the device to life:

> *The air pump was intended to produce an operational vacuum in its great glass receiver. By repeatedly drawing the piston (or "sucker") of the pump up and down and adjusting the value and stopcock connecting the receiver to the brass pumping apparatus, quantities of air could be removed from the receiver. The effort of drawing the sucker down became more and more difficult until it at last resisted all human effort. At that point Boyle judged that he had exhausted all atmospheric air from the receiver.*[21]

And the device let Boyle run experiment after experiment. His 1660 book *New Experiments Physico-mechanicall Touching the Spring*

---

* Hooke had what must have been one of the world's coolest roles, as in 1662, he served as the Royal Society's first "Curator of Experiments."

of Air and Its Effects* recounted forty-three separate experiments. He called the seventeenth the "principal fruit I promised myself from our engine."[22] He placed a device called a Torricellian apparatus, which used mercury to measure atmospheric pressure, into the receiver. Boyle expected that the level of mercury in the apparatus would fall as he removed air. It did.

Boyle was cautious about drawing conclusions from his experiments, with phrases like "perhaps it seems not impossible" in his writing. His caution shone through in the title of his next work, the 1661 book *The Sceptical Chymist: or Chymico-Physical Doubts & Paradoxes*, which took the form of a dialogue about the nature of matter. The book's discussion about how every phenomenon traces back to collisions of corpuscles in matter led many to consider his work the origins of modern chemistry.

More broadly, Boyle's application of Bacon's logic laid the foundation for generations of researchers. He was meticulous about documentation. His goal was to allow people to visualize his experiments and replicate them precisely. That meant documenting methods, materials, circumstances, locations, precise results, and the specific date and time when he ran the experiments. Since he viewed experimental research as a form of worship, Boyle often ran his experiments on Sundays. He dutifully reported failures as well as successes.*

Boyle largely withdrew from public life after suffering an illness in 1669. He died on December 31, 1691.

---

* I originally wrote here, "Given he was a fallible human." Reviewer Karl Ronn, who was a senior research and development leader in Procter & Gamble for two decades, commented, "Not sure human is the reason. This is standard practice in a bound journal. You don't know it is a failure until you fail. We used to have to record everything in journals in ink with countersigned pages to cover patentability when first to invent was the rule." A search of the US Patent and Trademark Office database shows forty-eight patent applications bearing Karl's name. His favorite? US patent 6,648,864, "Array of disposable absorbent article configurations and packaging." And now you know!

The Scientific Revolution's Echo: The Proclamation's Shadow

The king's proclamation feels like a historical relic. Modern organizations certainly don't go around telling people not to innovate. They don't lock them up for doing so. No ears get cut off. Rather, organizations seek innovation. They issue proclamations about innovation being a key to success in a world of never-ending change. A 2018 piece of research led by Professor Don Sull of the Massachusetts Institute of Technology that looked into stated corporate values of large companies found that the term *innovation* trailed only *integrity* and *respect* in frequency of use.

Talk to a modern executive, and they'll be quick to praise innovation. Yet most will admit that their organizations struggle with it. What's behind the barrier? I think the king's proclamation still echoes implicitly in people's heads. Disruptive innovation threatens the natural order of things. And who benefits from the current order of things? People who are currently in power.

Consider a conversation I had with a senior leader at a large, respected global company. I was working with a consulting team helping the company to develop a capability to more reliably create new disruptive businesses. As is done in any good consulting project, we started with a diagnosis of what was working well and what needed improvement.

"We are organized to deliver consistent, reliable results," the leader told us. This company indeed was well known for hitting its numbers and delivering reliable dividends to investors, who valued its stability. All good. "And that's exactly the problem," the executive added.

Hold on a second. Why is that a problem? When working on the book *Eat, Sleep, Innovate*, I had one of those duh moments about this conversation. Innovation is something different that creates value. Organizations exist, and are optimized, to do what they are doing better, faster, and cheaper. They seek to execute at ever-greater scale with ever-greater efficiency to produce ever-more-consistent results. That means prioritizing investments to make today better, not those that promise to make tomorrow different. Through this lens, innovations, particularly disruptive ones, can feel threatening. *That's* the problem.

Innovation is a break from the norm. It doesn't quite fit. It upsets the order of things. There's no standing Proclamation Against Those That Doeth Innouate, but many organizations in small, subtle ways *act* like there is.

The king's proclamation also echoes in how the world responds when new ideas emerge. In his landmark book *Diffusion of Innovation*, Everett Rodgers, a communications and journalism professor at the University of New Mexico, lays out an elegantly simple theory for how people adopt new ideas. He says that the population of would-be adopters was normally distributed. That means a curve with a distinctive bulge in the middle with three paired groups: about 2.5 percent at the left and right ends of the population, 13.5 percent approaching each side of the bulge, and 68 percent at the bulge. Rodgers names the left edge of the curve "innovators," people who would happily try just about anything. The next 13.5 percent he calls "early adopters," those who are more discerning but willing to try out early technologies with limitations. The next 34 percent (the first half of the bulge) was the "early majority," followed by another 34 percent of "late majority," and the final 13.5 and 2.5 percent "laggards." Published in 1962, the model is still in wide use today.

Humans suffer from what is known as the status quo effect. All things being equal, we like things to stay exactly the way they are.

Do you remember when you first saw Uber, or first considered using Airbnb for accommodations, or tried out ChatGPT? Maybe you are an innovator or early adopter who jumped in. It is more likely that you paused for a second. Novelty can feel a bit scary.

I remember getting in a Waymo robo-taxi in San Francisco in September 2024. The app provided a pickup point that required walking through a less-than-desirable side street. The car got stuck before it could turn, because people who may not have been in their right minds were loitering. When the car arrived, I nervously used the app to unlock the door. What sounded like dystopian music played, and a dull neon light glowed. I felt like I was at a bad club after everyone had left. The car set off. The route was circuitous, avoiding harder-to-drive-through parts of the city. At one point, we got stuck waiting to make a turn that I would have handled easily. My palms began to sweat.

And I make my living thinking about, writing about, teaching about how to navigate disruptive change! I'm still a human; I still have those feelings about things that seem to threaten the natural order of things.

When a business leader underfunds disruptive innovation or punishes rather than celebrates would-be disrupters that try and fail, when we pause before entering a robo-taxi or turning our life over to AI, we are all are acting in ways consistent with King Edward VI's wishes.

The proclamation is both a relic of its time and something that has a degree of timelessness to it. This stuff is hard.

The Responsibility of the Many

Clay Christensen originally described the phenomenon he had identified as disruptive technology. The first broad reference to disruptive innovation is a 2000 *Harvard Business Review* article, "Meeting the Challenge of Disruptive Change." The change in terminology is small but important. *Technology* makes us think about products. *Innovation* is a broad word, creating space for products, services, models, and, as this chapter shows, ideas.

In his scholarly review, Benoît Godin argues that the word *innovation* itself traces back to Greece in the fifth century BCE. The word *kainotomia*, derived from *kainos* (new), meant "cutting fresh into" or, in some literal uses "opening new mines."[23] Godin argued that innovation "acquired its current meaning as a metaphorical use of this word."[24] *Kainotomia* or its equivalent appeared episodically and generally had a negative connotation, as it went against natural order and stability. The word is largely absent from ancient Roman times, who would use *novitas* or *novatio* to talk about change. *Innovo* appears in the third or fourth century and is used as a form of renewal, often with religious connotations. Modern versions of the word *innovation* itself appear in France, England, and Italy in the thirteenth and fourteenth centuries. In 1658, the first edition of *The New World of English Words* defined innovation as "a making new, also a bringing in of new customs or opinions."[25]

Innovation is intensely human. Pick up an object in front of you. There are hundreds, maybe thousands, potentially millions, of people who played a role in its creation. Some helped build it, some developed the materials for it, some distributed it. Still others created knowledge that people used to build the machinery that produced it or created the trade laws that helped a global supply chain function. The names in this book, as in any book on such a complex topic, are really the tip of the iceberg. Regardless, nothing happens if humans don't do something.

Progress is not preordained. It takes forty-three experiments, conscious effort, work. There are failures along the way. Frustrations. Times when you can't see a path to success. There are people who never find the path to success, who never make it into the history books.

Bacon and Boyle made the books. And the tide turned. Seven years after Boyle's death, Thomas Savery patented a pump that was a precursor to the steam engine, which started the Industrial Revolution. The pace of change, which had been glacial for decades, began to accelerate. The American and French revolutions promised the possibility of a new way to govern and lead. And at long last, *innovation* became a positive phrase. Dictionary definitions shed negative connections to the word.

In 1782, English scholar Robert Robinson wrote, "Almost all great men that have appeared in the world have owed their reputation to their skill in innovating. Their names, their busts, their books, their elogiums, diffused through all countries, are a just reward of their innovations."[26]

Let's give Godin the final word on the topic before moving onto the nineteenth century to see disruption in health care:

> *The representation of being revolutionary, which had been negative until then, in turn gave innovation a positive meaning and gave a new life to the concept. Innovation acquired real political significance. While until then innovation had not been part of the vocabulary of innovators but rather a derogatory label and a linguistic weapon against innovators, it became a catchword in every discourse. Innovation gradually acquired a definite positive connotation due to its instrumental function. Innovation is a means to political, social and material ends. This culminated in the application of the concept to economic progress, through industrial or technological innovation in the twentieth century.*[27]

4

Florence from Florence

Revolutionizing Health Care
with Images and Words

The previous chapters described a disruptive product, platform, and idea. This chapter features a disruptive . . . pie chart?

I'm as surprised as you are. Those who have worked with me will report a few consistent quirks. If you send me something with two spaces after a period, you are likely to get a note about the beauty of proportional fonts. And I loathe pie charts, just loathe them! They require work to create, and even more work to process. You squint at the tiny font and try to compare pie slices of different sizes. A pie chart rarely serves as a compelling visual.

Almost never.

There's always an exception. And to be fair to the flames-on-the-side-of-my-face feeling pie charts generate, the chart featured in this chapter isn't *technically* a pie chart.* It is a polar area chart. André-Michel Guerry, a lawyer turned data analyst, created the first polar area chart in 1829, helping to usher in what one academic called the "golden age of statistical graphics."

"In the latter half of the 19th century, youthful enthusiasm matured, and a variety of developments in statistics, data collection and technology combined to produce a 'perfect storm' for data graphics," York University Professor Michael Friendly notes. "The result was a qualitatively distinct period which produced works of unparalleled beauty and scope, the likes of which would be hard to duplicate today."[1]

The polar area chart in question was published in 1858. It was visually stunning. More important, it highlighted a shocking reality: if you were a soldier in the early part of the Crimean War, you faced a greater chance of dying in an unhygienic hospital than you did on the battlefield. The statistical graphic had transformational impact.

The chart was the brainchild of arguably the first social media star of the mass communications era and a triple disrupter in health care: Florence Nightingale.

A Brief History of Florence Nightingale

Nightingale was born in 1820 in . . . Florence, Italy, naturally. She came from a well-to-do family. At seven she demonstrated an interest in the sciences, and at nine she began tracking the illnesses that afflicted her

---

* The "flames on the side of my face" refers to the 1985 movie *Clue*. I am a child of my time. *The Goonies. Ferris Bueller's Day Off. Dirty Dancing. The Princess Bride.* And, a few years after this run, *The Shawshank Redemption*. I've tried to get "get busy living or get busy dying" in three books, and it has been cut by editors three times. It can be in a footnote at least, right, Kevin? Right?

and her nine siblings. When her governess got married, her father, William, decided to personally educate his daughters in subjects such as Greek, Latin, physics, and chemistry.

The year 1837 was pivotal for the young Nightingale. Influenza broke out in her family's house in Hampshire, England, and on February 7, 1837, she said the voice of God called her to a life of service. Two years later, she was not materially closer to her goal, but she did get the chance to curtsy before Queen Victoria at the start of the debutante season. Later that year, she moved to London to help an aunt who was pregnant with her seventh child and to learn mathematics.

During the 1840s, Nightingale decided that the best way to answer God's calling was through nursing. Her family resisted. At the time, hospitals that would staff nurses were generally squalid places for the less fortunate, and the nurses were lower-class women with limited training seeking to earn extra income. She ended the decade rejecting a suitor, noting "to put it out of my power ever to be able to seize the chance of forming for myself a true and rich life would seem to me like suicide."[2] That year, while traveling with family friends in Europe, Nightingale managed to squeeze in a few weeks in a nursing institute in Kaiserwerth in Germany.

Two years later, her parents finally relented, and she went to Kaiserwerth for formal nursing education. In 1853, a family friend recommended that she serve as the superintendent at a charity hospital in London. She held that role for a year before resigning. She had hopes to become the superintendent of nurses at King's College Hospital in London, but an outbreak of cholera in the city and a conflict in Turkey stopped those plans.

The Horrors of Crimea

Nightingale's chart shows the causes of deaths during two periods of the Crimean War (a conflict between Russia and the allied England and France in the mid-1850s). The British Army hadn't fought in a war in forty years, and it was generally unprepared for battle. Poet Alfred, Lord Tennyson brought the battlefield brutally to life in his poem "The Charge of the Light Brigade," inspired by the October 1854 Battle of Balaclava. Published just six weeks after the battle, the second stanza rings through history:

> "Forward, the Light Brigade!"
> Was there a man dismayed?
> Not though the soldier knew
>   Someone had blundered.
>   Theirs not to make reply,
>   Theirs not to reason why,
>   Theirs but to do and die.
>   Into the valley of Death
>   Rode the six hundred.[3]

Secretary of War Sidney Herbert had asked Nightingale to bring a cadre of nurses to help with the increasingly dire situation. Conditions were not great for her to have impact: John Hall, the army's principal medical officer, held fast to traditional viewpoints. For example, he didn't believe in using chloroform, which was invented in 1831 and had emerged as the first meaningful anesthetic. "The smart of the knife is a powerful stimulant," Hall said, "and it is much better to hear a man bawl lustily than to see him sink silently into the grave."[4]

Nonetheless, Nightingale and her own brigade of thirty-eight nurses set sail, arriving at the war hospital in Scutari, which is now in the district of Üsküdar, Türkiye, on the Asian side of the Bosporus. From the outside, the hospital looked lovely. It was in a military barracks on cliffs that rose well above the water. The building was rectangular and big enough to fit eight professional football fields. There were stone towers in each of the hospital's seven corners.

On the inside, however... Patients were packed together on straw mattresses. Closed windows kept the cold air out but let the fetid air settle and recirculate. There were no clean clothes. No clean linens. No clean bandages. Rations had maggots. Mysterious substances made the water cloudy and led nurses to drink wine or beer instead. Rats raced all over the hospital. Soldiers' skin crawled with lice and fleas. For every forty patients with dysentery, there was *one* chamber pot.

It's hard to unhear the way physician Adam Rodman describes the hospital on his podcast *Bedside Rounds*: "In her private correspondence, Nightingale paints a graphic picture of a wave one-inch high of liquid feces flowing down the halls from the uncapped sewer, with the soldiers stricken with typhus and cholera walking through barefoot in order to relieve themselves."[5]

Through modern eyes, it is not surprising that rather than being a place of healing, the horrible conditions made it a tinderbox of communicable disease. Nightingale did what she could under the circumstances.

"The night scenes are quite Rembrandt," said one eyewitness. "Florence, the medical inspector general... and orderlies with candles surrounding a poor fellow on the ground with his arm off... Florence and Sister George binding up a stump, the surgeon on one side and orderly and light on the other."[6]

John Macdonald from the *London Times* ran with the image:

> *She is a 'ministering angel' without any exaggeration in these hospitals, and as her slender form glides quietly along each corridor every poor fellow's face softens with gratitude at the sight of her. When all the medical officers have retired for the night, and silence and darkness have settled down upon these miles of prostrate sick, she may be observed alone, with a little lamp in her hand, making her solitary rounds.*[7]

That image, a nurse with a lamp, stuck. The Crimean War was the first major conflict between great powers since the invention of the telegraph in 1837 had dramatically accelerated the pace at which communications could travel. The press provided day-by-day updates of the battle. And it made Florence Nightingale, the Lady with the Lamp, a megastar. The summer of 1855 was peak Nightingale mania. Before the war, there were about five thousand babies named Florence each year. After the war, they peaked at twenty thousand.[8]

In early 1855, Henry John Temple, Third Viscount Palmerston (known as Lord Palmerston), became the United Kingdom's prime minister. He sent a sanitary commission to Istanbul. Nightingale described how that team "cleaned sewers, unclogged water pipes, removed rotten floors, lidded latrines, and disinfected surfaces with lime wash; they removed 5,000 handcarts of filth, flushed sewers 477 times, and buried 35 animal carcasses. Monthly mortality rates of 23% in the first winter decreased to 2% in the second."[9]

Queen Victoria honored Nightingale with a brooch in early 1856. "I need hardly repeat to you how warm my admiration is for your services," the queen wrote, "which are fully equal to those of my dear and brave soldiers, whose sufferings you have had the privilege of alleviating in so merciful a manner."[10]

Upon returning to the United Kingdom, Nightingale visited the queen in Scotland, leaving with a charge to write a confidential report for her and a promise for the government to create a Royal Commission into hospital matters.

Shining a Light on a Deadly Killer

Nightingale came from the line of thinking that ran through Bacon and Boyle. She believed in capturing, categorizing, and analyzing data. She would later practice what we today would be called evidence-based medicine. For example, when data showed higher mortality rates in her own midwifery efforts, she stopped performing the services and suggested women instead deliver at home.

Nightingale also believed in the importance of communicating that data as clearly as possible. As she worked on her private report, she also emerged as a key background voice for the Royal Commission, which was run by former Secretary Herbert.

Nightingale connected with leaders in the emerging fields of statistics and visualization, most notably William Farr, a physician whom Nightingale had met at an 1856 dinner party. Born poor, Farr had a benefactor who supported his education and provided him with funds to become a doctor. While studying in Paris, he got a firsthand glimpse at the emerging work of Guerry, the visionary data analyst. Farr became an early member of the General Register Office, formed in 1836 to track births, deaths, and marriages in England and Wales. He sought to draw practical insights about mortality from necessarily imperfect data. He was part of the founding of the Statistical Society of London (later renamed the Royal Statistical Society) and served as its president from 1871 to 1873.

How could he, Nightingale, and others make the case that the military ought to make sure that a situation like Scutari never happened again, that small investments in the preventive value of sanitation and hygiene provided big payoffs by warding off communicable diseases and unnecessary death?

Data. And visuals.

The Statistical Society's logo was a bound sheaf of wheat, and its motto was *Aliis exterendum*, "For others to thresh out." The goal of a good statistician is not to advocate. It is to assemble the evidence to enable a robust, rational discussion about a topic.

Nightingale and Farr had a point of view, of course. And sometimes the power of data and visuals makes a point so obvious that it cannot be ignored. But the goal was truth. "We want facts," Nightingale wrote in a note to Farr. "'Facta, fata, facta' is the motto which ought to stand at the head of all statistical work. If we cannot have all the facts, let us on all events have all the reliable facts we can."[11]

Farr and Nightingale, with support from Farr's team at the General Register Office, developed a comprehensive database of military mortality. The work involved face-to-face meetings, furious correspondences by couriers, hand tabulations, and, perhaps, access to the office's Scheutz calculating engine, the world's second viable calculating machine. Editor and data storyteller R. J. Andrews details that machine vividly:

> *The machine worked by storing a series of numbers, each in their own geared column. These columns were mechanically linked so that turning a hand crank added particular columns in a specific order. The combined sequence of these simple operations allowed for complicated calculations to be performed, much easier than if they had been done by a human computer. The machine could handle fifteen-digit numbers, calculate four orders of differences, and output molds from which metal printing plates were cast.*[12]

Cool stuff.

Beyond the horrors of Scutari, the data showed that military men based in the United Kingdom, men who were safely away from combat and who should generally be healthier than the general population, had *lower* life expectancy than civilians.

Nightingale developed a powerful analogy to bring the data to life. "It is as criminal to have a mortality of 17, 19, & 20 per 1000 in the Line Artillery & Guards in England—when that of Civil Life in towns is only 11 per 1000—as it would be to take 11000 Men per annum out upon Salisbury plain & shoot them," she said.[13]

And the visuals. "Whenever I am infuriated," Nightingale wrote in a letter to Secretary Herbert, "I revenge myself with a new diagram."[14]

Farr and Nightingale created a raft of visuals. Bar charts illustrated this excess mortality by year. Stacked bar charts showed the number of deaths across soldiers, the general population, and the population in healthy districts. And, the polar area chart.

The chart appeared in what Herbert dubbed the "coxcomb," a short 1858 publication called "Mortality of the British Army, at Home, at Home and Abroad, and During the Russian War, as Compared with the Mortality of the Civil Population in England."* That publication primarily focused on visuals that otherwise might be lost in an eight-hundred-page official report. The title of the chart was "War in the East." An updated version (see figure 4-1) appeared in the 1859 publication *A Contribution to the Sanitary History of the British Army During the Late War with Russia*.[15]

The chart isn't simple. It demands attention. Where to look first? Naturally to the right-hand image, as it looks larger, more dramatic. You see something that looks like a clock with irregularly sized wedges for each hour-long interval. And rather than times, each interval shows a month. You see three shades, and the detailed description at the bottom explains that the lightest-shaded wedges are deaths from wounds, the medium-shaded areas are from preventable or mitigable "zymotic" diseases (e.g., fever, cholera, diarrhea, and tuberculosis), and the darkest-shaded wedges are deaths from all else.

The chart on the right covers April 1854 to March 1855, which includes the fateful Battle of Balaclava. Yikes. War is terrible, with many lost lives represented by the lightest-shaded wedges. But the medium-shaded wedge: Its size just jumps off the page. And then, you follow the line to the left side of the page, the later period of April 1855 to March 1856. You can't do the math to figure out exactly the difference, but it doesn't matter. The horrors of war continued, but the knock-on horrors of the hospital (the medium-shaded wedges) stopped.

* The era of brilliant titles seems to have passed, eh? Though the "coxcomb" is good.

Figure 4-1

Nightingale's polar area chart

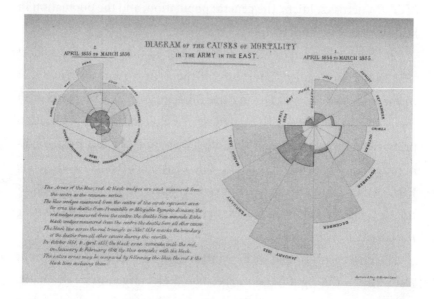

Source: Florence Nightingale, "Diagram of the Causes of Mortality in the Army in the East," in *Contribution to the Sanitary History of the British Army During the Late War with Russia* (London: John W. Parker and Son, 1859).

The text that accompanied the exhibits in the coxcomb brings the powerful visual to life:

> *The irregular blue surface, like the tail of the portentous comet, shows the zymotic diseases, the pests and scourges of camps and armies now, as they were of cities and towns in the middle ages, before the dawn of sanitary knowledge. . . .*
>
> *Compare then, the right with the left-hand diagram from the months of September to April, and no more instructive lesson on army hygiene could be given. The men were the same; the conditions only had been altered. The requirements of nature had been disobeyed in every particular during the first winter, and she has left on that diagram an everlasting vindication of her broken laws.*

During the second winter, nature had been more perfectly obeyed, and the stigma of her displeasure has almost ceased to appear....

The reduction in the mortality after the sanitary works were begun is most striking, and it falls eventually in June 1855 to less than a sixth part of what it was when the Barrack and General Hospitals were occupied together in October 1854, and to a nineteenth part of what it was in February 1855. Our General Hospitals have been so deplorably mis-managed in all our wars that the question has been raised as to whether it would not be better to do without them altogether. The experience of Scutari, as shown on the diagram, proves that General Hospitals may become pest-houses from neglect, or may be made as healthy as any other buildings....

The Crimean experience has proved the whole case, both as regards the disastrous results of defective administrative arrangements, and the possibility of foreseeing and obviating similar evils in the future.

And it has, moreover, shown that, with troops like ours, whose bravery and uncomplaining endurance of hardships the most severe and fatal, have been the admiration of the world, England has nothing to dread but the results of her own inexperience and want of foresight.[16]

In the words of Princeton University data scientist John Tukey, Nightingale's chart has "interocularity." The image hits you between the eyes, its message inescapable. "The graphic comparison is direct and immediate: far fewer died after than before," Friendly writes. "She was arguably the first to use this and other statistical graphs for political persuasion and popular impact."[17]

Nightingale's diagrams were well received in their time as well. "On every page, in every figure the enormous disproportion in the sanitary condition of the army is instantaneously and vividly impressed . . . the terrible results of carelessness and ignorance manifested so clearly and vividly in those awful diagrams," one newspaper reported. Here's another review: "Terrible do the death 'wedges' swell out during this period. Agonising to think how easily all might have been avoided, though there is consolation in the reflection that the woe may act as a warning."[18]

The work had lasting impact. The Royal Commission created committees to improve sanitary construction, build health codes, create a military medical school, and strengthen the gathering and interpreting of military statistics. In 1875, Britain passed the Public Health Act, calling for well-built sewers, clean running water, and regulated building codes. Over time, soldiers away from combat zones had better mortality statistics than the general population, which also saw its health shoot up in the decades ahead.

"She took data visualization, which is a rote, very detailed, technical graphic meant for reference and read only by super nerds who need to look at data," Andrews says, "and she transformed it into something that's more specific, which today we call data storytelling."[19]

Through the lenses of this book, even more so, Nightingale was a disruptive innovator. Disruption is about making the complicated simple and the expensive affordable.* In health care, that means allowing less specialized, less expensive providers to do more. From specialist surgeons to general practitioners to nurses to individuals. That approach brings care from expensive, centralized locations to less expensive, decentralized ones. From specialist hospitals to clinics to an individual's home. And the shift that Nightingale's work instigated: moving from treatment to wellness and prevention, which stop problems from occurring in the first place.

---

* Clay was passionate about disruption in health care. In 2000, he wrote a *Harvard Business Review* article titled "Will Disruptive Innovation Cure Health Care?" with physicians Richard Bohmer and John Kenagy. While working for Clay, I helped with the first draft of a book that expanded on the arguments in the article. Clay didn't think we had the argument quite right, so that book was shelved. Six years later, *The Innovator's Prescription*, by Clay and physicians Jerome Grossman and Jason Hwang, was published. The headline: Disruption could reframe the question from "How can we afford health care?" to "How can we make health care more affordable?"

Clearly, before Nightingale there were pioneering health-care innovators who developed surgical procedures, medicines, and an advanced understanding of disease. What is so striking about Nightingale—why I call her health care's first disrupter—is her ability to drive disruption in a very different way. I've detailed the first way she did so—her remarkable effort to shift focus from treating injuries to improving the conditions of hospitals to prevent infectious diseases. And she did it again, and again.

Enabling a Broader Population

While doing her work in Scutari, Nightingale contracted an illness, which historians believe was brucellosis, a bacterial infection that originates in animals. The illness left her with chronic symptoms, including nausea, headaches, joint pain, and extreme tiredness. While she lived until she was ninety, she was essentially an invalid for a large portion of the last fifty years of her life.

That didn't stop her from leaving her mark on the world. Her fame as the Lady with the Lamp led to donations pouring into the Nightingale Fund. She used that to help open a nursing school at St Thomas' Hospital in London. The first batch of students arrived in 1860. Graduates from that program, known as Nightingales, spread across the world, helping to professionalize nursing. This is the second disruption, enabling a broader population to practice nursing. I'm struck even more by something Nightingale did in parallel that had arguably even deeper impact.

She wrote.

"The pen may not be mightier than the sword," organizational psychologist Adam Grant likes to say, "but its ink lasts longer."[20]

Disruption takes things that are complicated and expensive and makes them simple, accessible, affordable, and able to reach greater populations. Nightingale's diagrams literally hit you between the eyes.

If you have disruptive innovation on your mind, reading *Notes on Nursing: What It Is and What It Is Not*, Nightingale's book on nursing, is just as interocular.

"At the time it was first published, there were no schools for nurses and therefore no trained nurses," writes Virginia Dunbar, former dean of the Cornell University School of Nursing, in the foreword to the 1969 republication of the book. "But more important, there was no body of knowledge, no recognized sphere of activities which constituted nursing either as it should be given by the mother, or as it might form the basis for preparing a special group to be known specifically as nurses."[21]

The book wasn't meant for people seeking a career in nursing. It was meant for the everyday person who might sometime have to provide the same duties as a nurse. Its original print run of fifteen thousand copies sold out within months of its 1859 launch. An American version came out in 1860. The book was quickly translated into multiple languages. A simplified version with technical language stripped out—*Nursing for the Labouring Classes*—came out in 1861.

The book is readable and highly practical. Nightingale focuses on the importance of spending time with patients, providing clean sheets, fresh air, good food, hygiene, and access to sunlight. "In surgical wards," she writes, "one duty of every nurse is prevention. . . . The surgical nurse must be ever on the watch, ever on her guard, against want of cleanliness, foul air, want of light, and of warmth."[22] Running through the book is Nightingale's belief about the importance of good observation:

> *In dwelling upon the vital importance of sound observation, it must never be lost sight of what observation is for. It is not for the sake of piling up miscellaneous information or curious facts, but for the sake of saving life and increasing health and comfort. The caution may seem useless, but it is quite surprising how many*

> men (some women do it too), practically behave as if the scientific end were the only one in view, or as if the sick body were but a reservoir for stowing medicines into, and the surgical disease only a curious case the sufferer has made for the attendant's special information.[23]

My hatred of pie charts? Nothing compared to Nightingale's withering commentary on the uselessness of most dusting techniques. "No particle of dust is ever or can ever be removed or really got rid of by the present system of dusting," she writes. "Dusting in these days means nothing but flapping the dust from one part of a room on to another with doors and windows closed. What you do it for I cannot think. You had much better leave the dust alone, if you are not going to take it away altogether."[24] She goes on to illustrate her practicality, describing how the best wall surface to combat dust is oil paint, why the best floor is a Berlin lacquered floor, and how furniture should be made of material that can be wiped clean with a damp cloth.

Florence Nightingale died in her home in London on August 13, 1910. She was ninety years old. Five years later, during the brutality of World War I, Arthur George Walker unveiled a statue of Nightingale in Waterloo Place.* She has a faint smile and is, of course, holding a lamp.

Florence Nightingale's Echo: Freedom House Paramedics

The Freedom House story simultaneously reflects on the lessons from Nightingale's story and echoes to modern times. Call it a reflective echo.[25]

---

* I worked less than a quarter mile from this statue in 1997. Small world.

Before the 1960s, if you suffered a health emergency and needed to get to a hospital, depending on where you were, you might be transported in a fire truck, a police car . . . or a hearse (not because you were dead, but because a funeral house's hearse was an underutilized vehicle). The goal was to get you to the hospital as quickly as possible. In 1966, the National Academy of Sciences published a report titled "Accidental Death and Disability: The Neglected Disease of Modern Society."[26] The report ended up being so influential, today emergency medical technicians simply call it "The White Paper."

The report described how as many as fifty thousand annual deaths could be attributed to the lack of proper prehospital care:

> *One of the serious problems today in both the lay and the professional areas of responsibility for total care is the broad gap between knowledge and its application. Expert consultants returning from both Korea and Vietnam have publicly asserted that, if seriously wounded, their chances of survival would be better in the zone of combat than on the average city street. Excellence of initial first aid, efficiency of transportation, and energetic treatment of military casualties have proved to be major factors in the progressive decrease in death rates of battle casualties reaching medical facilities.*[27]

That same year, an intriguing solution emerged in the Hill District, a predominately Black part of Pittsburgh. A Black-operated jobs training program called Freedom House had rolled out a mobile service that would transport fresh vegetables to people's doors. Phil Hallen, a former ambulance driver who headed a civil rights organization, connected with Peter Safar, the head of anesthesiology at the University of Pittsburgh. Safar was known as the "father of CPR (cardiopulmonary resuscitation)" for his work in developing standard, simple ways to train broader populations in the lifesaving resuscitation technique.*

"You are just what I'm looking for," Hallen recalls Safar exclaiming when they met. "I've been trying to figure out how to take rescue breathing and the rest out on the street and to train people how to do that. You've got the people. . . . Let's train them to be professionals."[28]

"It was not, 'go to the scene, pick up a patient, transport to a hospital, and then start care,'" Hallen says. "It was emergency treatment right there on the pavement."[29]

The two developed an idea to train nonmedical professionals to provide basic treatment on-site, with special-purpose ambulances that essentially were mobile intensive care units. It is a standard idea today, but nothing becomes standard unless someone has the flash of insight, turns insight into a concrete plan, and acts. And that's what Safar and Hallen did.

Safar created a three-hundred-hour course that taught the first batch of first responders everything from advanced first aid to defensive driving. They learned techniques such as IV insertion, intubation, and spinal immobilization. And then they applied those techniques in the field. Soon, Freedom House's fully Black team was taking six thousand calls a year. There were memorable moments, such as when a Black boy riding a bike in one of the city's most affluent neighborhoods had a head-on collision with a car. The police arrived but didn't know what to do, so they called Freedom House, whose responders saved the boy's life. Safar's data showed that the paramedics saved two hundred lives in Freedom House's first year.

---

* A good historical footnote: To demonstrate the power of CPR, Safar ran an experiment where he paralyzed a group of volunteers using a substance normally used on poison-tipped arrows in the Amazon. He then had a bunch of children administer CPR to keep the volunteers alive. Thankfully, the experiment worked.

In 1975, the federal government asked Nancy Caroline, Freedom House's medical director, to write the first standardized training curriculum for emergency medical service providers. *Emergency Care in the Streets* followed in Nightingale's footsteps, enabling a broader population to learn about providing care in a decentralized setting.

That same year, Pittsburgh Mayor Pete Flaherty announced plans for the city to roll out its own paramedic service. Although many of Freedom House's staff joined the program, within a few years they were reassigned, and as late as the 1990s, some 98 percent of Pittsburgh's paramedics were white.

Today, there are close to two hundred thousand paramedics in the United States, saving lives by providing care in decentralized settings.

Disruptive innovation does not depend on technological breakthroughs. There are multiple ways to transform what exists and to create what doesn't, and people with diverse backgrounds and skills can drive disruption. So, the next time you see a nurse automatically sanitize their hands as they enter a room, think of the Lady with the Lamp and how she changed the world with powerful charts and simplifying language.

And remember that innovation is not the job of the few. It is the responsibility of the many.

5

The Great Democratization

How the Model T Put the World on Wheels

In the 1920s, a fierce battle broke out for the soul of city streets.

The idea of mechanizing personal transport had long been intriguing but also polarizing. Country roads were built for horse-drawn transport; urban roads were primarily for pedestrians. When Roy Chapin drove an Oldsmobile runabout from Detroit to New York in late 1901 at the heady speed of ten miles an hour, mule drivers constantly harassed him by yelling, "Get a horse!"[1]

Should streets be primarily for people or for cars? It wasn't an academic battle. Henry Ford's Model T, the disruptive innovation on which this chapter focuses, had dramatically increased automobile adoption. That spurred chaos and carnage on city streets. Newspaper cartoons in the 1920s often showed the Grim Reaper driving cars. One editorial drawing in the *St. Louis Star* showed a man kneeling and holding up a platter of children to a car with a humanoid, maniacal grin. The cartoon's title? "Sacrifices to the Modern Moloch," referencing a god of fire only appeased by child sacrifice.[2] In 1922, the mayor of Baltimore dedicated a twenty-five-foot wood and plaster obelisk as a monument for the 130 children who had died in motor accidents that year.

Resolving the battle involved developing traffic signals and norms about rights-of-way. It also involved dueling public relation campaigns.

Motorists coined a term for pedestrians who crossed roads in the wrong places or in the wrong ways. The word *jay* then meant an uneducated country dweller who was out of place in the city. Thus, the epithet *jaywalker*. Motorists enlisted Boy Scouts to distribute anti-jaywalking cards to pedestrians. Pedestrians fought back. They sought to brand drivers who drove recklessly as *flivverboobs*.*

By 1924, the magistrate of the New York City's Traffic Court ascribed somewhere between 70 and 90 percent of accidents to jaywalking.[3] If you type *flivverboob* into Microsoft Word, the distinctive red squiggle shows that the word is no longer in use. In fact, it had largely disappeared by the late 1920s.

Public sentiment shifting toward motorists was more than a marketing victory for automobile enthusiasts. It was a key step in the transformation of transportation. By making an expensive, complicated product affordable and relatively easy to use, Henry Ford played a critical role in creating an industry that helped create the modern world. The story begins in the late nineteenth century in Dearborn, Michigan. Let's roll.**

---

* Yes, *flivverboob* has two *v*'s. Would history have turned out differently if it had only one?

** There are going to be a lot of bad driving puns in this chapter. That's just the way it is going to be.

A Brief History of the Model T

Henry was born in Dearborn in 1863.* It was a frontier town that shifted from agriculture to industry during his life. He grew up on a farm and would often extol the virtue of good, honest work. In 1876, the same year his mother died, he saw a vehicle powered by a steam engine, which one of the many biographies of Henry described as "a sight almost as astounding to the boy as if Elijah's chariot of fire had suddenly appeared."[4] Donn Werling, the former historical director of the Henry Ford Estate, had similarly evocative language, calling it "an earth-shattering experience. With its noise and smell, the only similar image would be a large cannon. But instead of exploding bombs, the machine exploded a way of life."[5]

Henry was a tinkerer by nature. He was particularly fascinated by watches and enjoyed taking them apart and putting them back together. Ford R. Bryan, who, remarkably, wrote five books on Henry Ford, said, "Henry would do anything to get out of his chores. . . . He'd pester people to loan him their watches so he could figure out how they worked to the point where the whole neighborhood said, 'You'd better keep an eye out for Henry Ford or he'll take your watch apart.'"[6]

In 1891, Henry got a job at Detroit's Edison Illuminating Company with a lofty salary of $40 a month. By day, he would help part of Thomas Edison's company bring light to the masses; at night, he would disappear into his shed and keep tinkering away. "I cannot say it was hard work," Henry later noted. "No work with interest is ever hard."[7]

Europeans had the early lead in key automobile technologies. In 1859, Jean-Joseph Étienne Lenoir of Belgium invented the internal combustion engine. In 1864, Siegfried Marcus of Vienna built a gas-powered

---

* It breaks convention, but this chapter will use Henry or Henry Ford to refer to the person, and Ford as shorthand for the Ford Motor Company. It just felt right.

road vehicle. In 1876, Nikolaus Otto of Germany built a four-stroke internal combustion engine. American ingenuity appeared two decades later. In 1896, Ransom E. Olds built his first gas-powered car, in Lansing, Michigan. The same year that Olds built his first car, Henry built *his* first functional automobile, the *quadricycle*.

The year Henry built his quadricycle, he met Edison, who allegedly told him, "You have it—a self-contained unit that carries its own fuel. That's the thing! Keep at it!"[8]

Henry did just that. But he had to play catch-up. In 1897, Olds founded Oldsmobile. The company sold four hundred cars in 1901, twenty-five hundred in 1902, four thousand in 1903, and fifty-five hundred in 1904.[9] Those figures were meaningful in the emerging automobile category, but they were tiny compared with the roughly sixty thousand horse-drawn carriages sold annually in New York City alone.[10] In 1904, investors forced Olds out of his namesake company. A few years later, General Motors acquired Oldsmobile.

Meanwhile, Henry was balancing technical progress with commercial challenges.

He founded the Detroit Automotive Company in 1899, with the backing of twelve investors. The company blew through $86,000 of financing (roughly $3 million today) to produce twenty cars. It folded in 1901. Henry then formed the Henry Ford Company, with William Murphy and Lemuel Bowen serving as financial backers.

Henry's focus was turning his quadricycle into a functioning automobile. And he created a good one. He sought public ways to demonstrate its prowess. On October 10, 1901, Henry himself sat in a car named "Sweepstakes" to race Alexander Winton of the Cleveland-based Winton Motor Carriage Company. Winton had race experience, and his car, "Bullet," was significantly more powerful than Henry's car.* Because

---

* History is mixed on how much more powerful Winton's car was. Henry's car apparently had 25 horsepower; different accounts had Winton's at either 45 or 70 horsepower.

the race was in Grosse Pointe, Michigan, Henry was the eight-thousand-person crowd's hometown favorite. Winton got out to an early lead, but on the seventh of ten laps, his car slowed with smoke streaming out of the engine. Henry sped past and won the race.

Henry never raced again. Perhaps his backers weren't happy with how much time he was spending on the racetrack, viewing it as a distraction. In 1902, he left the company that bore his name. Before liquidating the company, his financial backers asked Henry M. Leland to appraise their factory and tooling. Leland had designed the engine for the Oldsmobile. He suggested that the company combine his engine with its existing chassis and remain in business. The backers agreed, hiring Leland and renaming the company Cadillac. General Motors then purchased Cadillac for $4.75 million in 1909.

These short vignettes show a nascent industry replete with uncertainty. Because automobiles were complicated and expensive, the market was limited. The opportunity was there for a company to simplify, to standardize, to disrupt.

The company that did just that was formed in 1903, when Henry and eleven associates banded together to create his third (and final) company: Ford Motor Company.

"A Car for the Great Multitude"

Henry Ford 1.0 tinkered in his garage. Henry Ford 2.0 developed an impressive machine but might have been distracted by the allure of the racetrack. Henry Ford 3.0 had a clear vision: make a car accessible to a broader population. "I will build a car for the great multitude," he said, "constructed of the best materials by the best men to be hired after the simplest designs that modern engineering can devise. But it will be so low in price that no man making a good salary will be unable to own one—and enjoy with his family the blessing of hours of pleasure in God's great open spaces."[11]

Henry Ford had a clear vision. But how to make it real? That wasn't clear, at least at first. Charles E. Sorenson, the company's production chief, said, "Ford merely had the idea; he had no picture in his mind as to what the car would be like, or look like."[12]

Henry liked his letters. In 1903, his company introduced the Model A, an eight-horsepower, two-cylinder vehicle with twenty-eight-inch wheels with wooden spokes and three-inch tires. Advertisements called it "positively the most perfect machine on the market." It cost $850, about $30,000 in today's dollars. Ford sold about two thousand Model As over the next couple of years.[13]

Henry struggled to sell his vision to his backers. Part of the problem was he hadn't *delivered* on that vision. You can feel the tension in the high-priced models Ford introduced in the next few years. The four-cylinder Model B sold for $2,000. The Model K sold for $2,800.

What would have happened had Henry abandoned his vision in favor of a luxury path? Perhaps he would have gone down in history as a footnote while someone else put the pieces together to blow open the market. The Model K disappointed. It sold fewer than a thousand units. Vindicated, Henry bought out the shareholders that were pushing in the premium direction, consolidated his control over the company, and accelerated the path toward simplification.

The Model N served as a tipping point. Introduced in 1906, it was marketed for $600, or $20,000 in today's terms. The car had a lighter, more powerful engine and a simplified design. It sold almost seven thousand units between 1906 and 1908. "The Model N was the direction Henry wanted to go," says transportation historian Bob Casey. "It was small, durable, light, and a great value for the money. But it wasn't big enough to squire four or five people; it hauled three, at best. In a way, Henry was still practicing to get it right, tinkering to find the perfect car in subsequent models."[14]

We're now in 1908. A transformative year for Ford, and for the industry.

Ford had sold about twenty thousand cars as it went through its various letter models. In October, it rolled out the Model T. It was a marvel.

Ford partnered with J. Kent Smith, an English metallurgist, to use vanadium steel in the car's crankshaft and chassis. This innovative metal reduced the weight of the vehicle, enabling more-elegant designs and better performance. The four-cylinder engine was a single block, as opposed to four elements that were bolted to separate crankcases. Creating the single-block engine was a significant challenge, but the result was an engine that was simultaneously stronger, lighter, and cheaper. Henry replaced dry batteries with a flywheel magneto that created a current that ingeniously delivered electricity to the engine. The car was the first to feature left-hand drive, and it had significantly lighter clutches than in existing cars and arc springs over the axles to help reduce painful jerks on bumpy roads.

The Model T was light. It was powerful. It was sturdy. It was different. And it was, relatively at least, cheap. The car debuted at $850 ($29,250 today), a premium to the Model N but a sharp discount to the Model K. That was just the beginning. Incremental improvements lowered the price to $590 ($20,000) by 1911. Henry had the foundation.

To deliver a car for the multitudes, however, he needed to figure out more-inventive ways to dramatically lower prices.

An Innovative Way to Lower Prices

If you play word association with the name Henry Ford, most people will respond not with "Model T" but with "mass production" or "assembly line." Indeed, when the Ford Motor Company perfected its assembly line, it dramatically increased the productivity of production, allowing Henry to dramatically lower the cost of his car.

Workers assembled the first Model Ts at a Detroit factory on the corner of Beaubien Street and Piquette Avenue before moving to nearby

Highland Park. It took skilled workers about 12.5 hours to create a car. The need to speed up production was clear. In 1910, J. S. Seaton wrote in *Harper's Weekly,* "The man who can successfully solve this knotty question and produce a car that will be entirely sufficient mechanically, and whose price will be within the reach of millions who cannot yet afford automobiles, will not only grow rich but be considered a public benefactor."[15]

Henry didn't develop the assembly line and reduce the price of the car. His obsession, clearly explained in his 1923 autobiography, was to make his product affordable and accessible:

> *Our policy is to reduce the price, extend the operations, and improve the article. You will notice that the reduction of price comes first. We have never considered any costs as fixed. Therefore we first reduce the price to the point where we believe more sales will result. Then we go ahead and try to make the prices. We do not bother about the costs. The new price forces the costs down. The more usual way is to take the costs and then determine the price; and although that method may be scientific in the narrow sense, it is not scientific in the broad sense, because what earthly use is it to know the cost if it tells you that you cannot manufacture at a price at which the article can be sold? But more to the point is the fact that, although one may calculate what a cost is, and of course all of our costs are carefully calculated, no one knows what a cost ought to be. One of the ways of discovering . . . is to name a price so low as to force everybody in the place to the highest point of efficiency. The low price makes everybody dig for profits. We make more discoveries concerning manufacturing and selling under this forced method than by any method of leisurely investigation.*[16]

Figure 5-1

Ford's assembly line in action

Source: Reprinted with permission of Fordimages.

The "highest point of efficiency" required radically rethinking production. The basic idea of the assembly line is obvious in retrospect. Instead of bringing workers to machines, bring the machines to individuals (figure 5-1). In 1904, rival Ransom E. Olds started experimenting with an idea to place vehicles onto wooden platforms supported by rolling casters, allowing stationary workers to attach components as the in-production vehicle rolled. Olds experimented. Henry nailed it.

Like Gutenberg's printing press, Henry's assembly line combined approaches from other industries such as meatpacking and sewing.

The first experiments mirrored Olds. Workers would place cars on movable benches and push them from one station to another. Henry then tried using inclined slides to move parts from one production station to another. In April 1913, Henry broke the production process into twenty-nine steps. In October 1913, workers used a rope and windlass to pull a chassis across the factory floor. One hundred and forty workers manned the 150-foot production line. Production time declined from 12 hours and 30 minutes to 5 hours and 50 minutes. That time declined to a mere 93 minutes in 1914.

Production exploded. In 1912, the company produced 82,388 Model Ts. In 1914, it produced 308,162, more than the rest of the industry combined. By 1917, the number had more than doubled to more than 700,000. Prices dropped as well. In 1916, the Model T sold for $360 ($10,500 today). In 1924, it sold for $260 ($5,000). It was, at last, a car for the multitudes.

Henry kept pushing. He imagined using the two thousand acres he owned on River Rouge to create a mammoth factory where blast furnaces and coke ovens turned raw materials into finished parts and where production lines connected to railroad lines and a dock to speed distribution. By the time the location was operational, it had its own power plant that produced enough electricity to power the city of Boston, and the Ford Motor Company established its own lumber operation (in Michigan) and its own rubber plantation (in Brazil) for tires.

We would now call this kind of end-to-end system *vertical integration*. Mastering mass production allowed Ford to operate at a scale that was previously unimaginable. All in service of lowering prices and making his car more affordable for the masses.

There were no robots in the 1910s and 1920s. Another piece of the disruptive puzzle was solving the pressing problem of high worker turnover. Henry had a simple answer. Pay them more. A *lot* more.

An Innovative Way to Lower Turnover

January 6, 1914. The headline stretched across the entire front page of the late edition of the *Detroit Journal*.* It was two lines, in fact, taking up about 10 percent of the page. It was in italic bold uppercase font. "HENRY FORD GIVES $10,000,000 IN 1914 PROFITS TO HIS EMPLOYEEs [*sic*]." The article's subhead, in slightly smaller font on the right side of the page, read, "DOUBLES PAY OF 25,000 IN AUTO WORKS."

The move shocked the world. At the time, the average worker received about $1.75 a day (about $55 today, or less than $7 an hour, assuming an eight-hour day).** Turnover plagued many businesses. In 1913, Ford offered Christmas bonuses to workers who had been with the company for more than three years. Only 640 workers qualified. But on that day in January, he announced that henceforth Ford workers would receive $5 a day (about $160 today, or $20 an hour).

Was Henry a kindhearted soul seeking to offer more to his workers? Perhaps. He regularly extolled the virtue of putting purpose before profits, noting that "a business that makes too much profit disappears almost as quickly as one that operates at a loss."[17]

Was he seeking to create more customers? Perhaps. That is certainly what he proclaimed. But with a car that still cost about seven months of wages, the claim seems somewhat dubious.

Was he a coldhearted tycoon seeking to find ways to further optimize his model and tie workers to his increasingly efficient factory?

---

* What's the late edition, you wonder? Believe it or not, some newspapers used to print multiple editions in a day. The late edition could be either the one that went to the press just before the crack of dawn or the one created at the end of the day. Gutenberg's gift created this kind of possibility; the disruptive power of distributing content over the internet obviated it.
** They certainly didn't receive the kinds of benefits that many workers take for granted today.

Perhaps. A wife of one of his workers thought so, writing in a letter, "The chain system you have is a slave driver Mr. Ford. My husband has come home and thrown himself down and won't eat his supper—so done out. Can't it be remedied? . . . That $5 a day is a blessing—a bigger one than you know—but, oh, they earn it."[18] Henry himself noted that the wage increase was the best cost-cutting move he had ever made. In 1914, attrition meant the company hired fifty-three thousand people annually to keep its workforce constant. In 1915, that number dropped by 80 percent to sixty-five hundred.[19]

No one can know his rationale for sure. It certainly captured attention, with the *Wall Street Journal* calling the move "economic blunders if not crimes." Whatever your interpretation, it clearly ended up providing further fuel for a powerful, reinforcing model that drove costs relentlessly down.

Driving Democratization

The Model T's price made it affordable. The final piece of the disruptive puzzle was making sure it also was *usable*.

In his classic article "Marketing Myopia," Theodore Levitt, the guru of marketing at Harvard Business School in the 1960s and 1970s, called Henry "both the most brilliant and the most senseless marketer in American history."[20] Even his senselessness—famously telling customers they could have "any color they wanted as long as it was black"—had its moment of brilliance, because it allowed standardization, simplification, and low prices.

What made Henry singularly brilliant was the degree to which he based his business around the customer. It wasn't just about price. It was about making the car usable to broader populations.

Would-be motorists faced more problems than being branded a flivverboob: Cars broke down, and there was not a wide infrastructure of service stations to help in an emergency. Ford's cars were complicated to build but simple and dependable to operate. The public nicknamed the car Tin Lizzie, which at least some historians connect to slang for a dependable servant. Ford ran public clinics to show how to fix common problems. And for $1.50 (about $25 today), you could purchase a kit that had two headlamp bulbs, a taillamp bulb, a spark plug, and a tire repair kit. And the layperson could easily fix their own car.

That's an important part of the Model T story—allowing nonexpert mechanics to confidently run an automobile. Consider the story of a Model T driver who drove with his wife Ethel, their infant daughter, and his sister-in-law from Chicago to Florida in the 1920s.

> *Ethel and I stored our furniture, cranked up the Model T, and headed south on the old Dixie Highway. It was a memorable trip. I had five new tires when we left Chicago. When we arrived in Miami, not one of those originals was left on the car. It seemed like we averaged a blowout every fifteen or twenty miles. I'd jack up the car and pull the wheel to patch the traitorous innertube and sometimes while I was applying the glue or manning the air pump, another tire would go bang! and expire. The roads were pretty primitive, of course, especially those red clay tracks through Georgia. At one point we came to a washout where the road disappeared and was replaced by a hog wallow. Ethel held the baby in her lap and steered the car while her sister and I pushed, sinking knee deep in the red muck.*[21]

We'll meet this driver, Ray Kroc, in chapter 8 as we see how he brought fast-food to the multitudes with McDonald's.

The personal part of the Henry Ford story ends in 1947. On April 7, a storm caused flooding in Henry Ford's local area. Biographer Bryan explains what happened next: "Henry went out to examine the damage, walking with his driver of the wet ground in his bedroom slippers—his 'clodhoppers' he called them—to the powerhouse."[22] Henry slipped, injured himself, and passed away later that day.

The world paused to remember a remarkable person. Detroit, whose population had surged from 305,000 in 1900 to close to 2 million in 1947, displayed a portrait of Henry Ford in front of city hall for thirty days. The day he was buried, all cars in Detroit came to a complete halt as Henry's body was lowered into the grave. *All* the assembly lines in the United States stopped for one minute. As Harvard historian Richard Tedlow notes, "Nothing like this had ever accompanied the death of an American businessman in the past. Nothing ever will again. An appropriate gesture for the Copernicus of cars."[23]

Anomalies by Anomalies

Henry Ford liked to fix watches. So, let's use his story to reflect on what makes disruptive innovators tick.

Henry was a man of contradictions. Paradoxes even. He grew up on a farm but ushered in the urban era. He proclaimed himself a modest man, but he became one of the world's wealthiest individuals. He loved to tinker and work individually, but his company operated at a pace and scale that was previously unimaginable. His wage increase simultaneously freed and bound workers. He was a craftsman whose assembly line destroyed craftwork.

Full biographies (the honest ones, at least) explore other contradictions. In 1915–1916, he took an overseas trip in a quixotic effort to stop the world war. A pacifist? A year later, his factories were churning out equipment for the US military, with Henry noting, "I am a pacifist first but perhaps militarism can be crushed only with militarism. In that case, I am on it to the finish."[24] A patriot? Some of both?

A complex man? A simple man? An anti-Semite? In 1918, he purchased the *Dearborn Independent*, which unquestionably started an antisemitic campaign in 1920. Some of the articles in that campaign found their way into a volume called *The International Jew*, which one Nazi leader said, "You have no idea what a great influence this book had on the thinking of German youth."[25] In a 1927 suit, Henry disavowed direct knowledge of the articles, settled out of court, and printed an apology in the *Dearborn Independent*.

Tedlow's book that profiles Henry Ford and other giants of enterprise describes how, by definition, great businesspeople break rules: "Rules are often little more than the unconscious assumptions of unimaginative people. When rules go beyond conventional thinking and are codified as laws, those laws are invariably written for the benefit of people who have done well in the past and are doing well in the present. They are not written to enable breakthrough thinkers to turn the world upside down."[26]

Clay used to have a sign in his office that read "Anomalies Wanted." Like a good social scientist, he believed that the things you didn't expect provided the richest learning. A disruptive change is an anomaly, something unexpected that has cascading, often unanticipated effects. It is perhaps not a surprise that the protagonists themselves are anomalies.

The Model T's Echo: Lab-Grown Meat and Artificial Intelligence

Ford introduced a stream of innovations that dramatically accelerated the growth of the automobile industry. The combination of assembly-line techniques, scale economics, and compensation that lowered worker attrition allowed Henry to truly live up to his promise of creating a car for the great multitude. By transforming an expensive, complicated product into an affordable and relatively easy to use one, he played a critical role in shaping an industry that helped create the modern world.

It is a classic disruptive innovation.

The Model T story took decades to unfold. It took place at the hyper-specific level of an assembly-line worker doing *this* instead of *that* and at a system level where broad pieces eventually clicked together in reinforcing ways. It involved innovations such as novel technologies and collective efforts to enable the automobile's growth by stigmatizing pedestrians.

In these complications, we see the opportunities and challenges of disruptive change. The opportunity to transform what exists and create what doesn't presents the tantalizing possibility of changing market dynamics, and indeed, changing the world. The challenge is managing a complicated, multifaceted effort to build a system supporting the disruption. And to do that in a world where the disruption is not a universal good.

The battle between jaywalkers and flivverboobs repeats throughout history. You see it in King Edward VI's proclamation. It's in the bands of English workers (branded "Luddites"), who in the 1810s destroyed machinery that they thought would threaten jobs in cotton and woolen mills.

The shadow cast by disruption is readily apparent in modern innovations as well. Consider lab-grown meat. It offers the prospect of simple, affordable solutions that could dramatically change the dynamics of the food industry. This innovative product also offers substantial environmental benefits, as raising cows requires significant amounts of land and the cows themselves produce methane. As lab-grown meat went from a science project to a viable commercial proposition in the 2020s, naysayers began to dig in their heels. Locations with high concentrations of ranchers, farmers, and food processors explored legislative changes to inhibit industry growth. Propaganda campaigns recalled the battle between flivverboobs and jaywalkers.

Another example is generative artificial intelligence (gen AI). The 2022 release of OpenAI's ChatGPT brought the technology to the mainstream. Early adopters marveled at the fluid prose, compelling synthesis, and occasionally humorous joke churned out in milliseconds. Further advancements led pundits to herald the technology's transformational potential. And yet . . . there's another side, a shadow.

In the spring of 2024, I taught an experimental class about how to use gen AI to aid in decision-making. The MBA classroom was a safe zone. The class was graded pass/fail. Nevertheless, after the class ended, many students reflected on how fearful they were about using gen AI for tasks such as developing creative ideas or creating compelling ways to communicate those ideas. People like change that's good for them. But change that requires them to unlearn old habits and learn ones, they like less. And change that potentially threatens their very identity, they like even less. Something about gen AI's humanlike nature triggered that kind of fear.

The first time a generation falls under the shadow of a disruptive change, these kinds of stresses and fears feel new, but any big shift creates friction and division. Disruption casts a shadow.

6

Another Big Bang

The Transistor's Unexpected
Path to Transformation

In 1948, three scientists posed for a picture to mark a discovery that would change the world. The picture would be blasted around the world in 1956, when the three shared the Nobel Prize. Can you spot the simmering tension in figure 6-1?

It would be reasonable to assume that John Bardeen (standing, wearing glasses) was, in modern language, throwing shade at Walter Brattain (also standing) and his research director, William Shockley (seated). Nope. Bardeen and Brattain had partnered in late 1947 to successfully run a proof-of-concept test of what became known as the transistor. Brattain ended up ruing some of the knock-on effects of his invention, most notably the unexpected boost the transistor gave to the emergence of rock and roll. By all accounts, however, the two got along swimmingly.

Figure 6-1

The Nobel Prize–winning inventors of the transistor

From left, John Bardeen, William Shockley, and Walter Brattain.
Source: *Bell Telephone Magazine*, August 1953.

Despite his faint grin, Shockley was the one who felt deep unease. Although he oversaw the research effort that spurred the development of the transistor, he had little to do with the critical breakthroughs that led to the first successful test. His exclusion from Bardeen and Brattain's breakthroughs frustrated him. Shockley was a mercurial character who liked Hollywood and fast cars and who left his wife after she was diagnosed with cancer. A combination of jealousy and determination drove him to launch a splinter research effort to produce an alternative technology that hastened the transistor's commercialization. He left the research labs in the 1950s and played a pivotal role in forming both the modern semiconductor and venture capital industries before taking a dark turn toward eugenics and dying in disgrace.

The goal of the three scientists was to develop technology that would improve the performance of telecommunications networks. While the technology ultimately achieved that aim, the path the transistor followed had the unexpected secondary effect of enabling the miniaturization of electronics and ushering in the modern information era. This development enabled the radical reconfiguration of multiple industries—a gale of creative destruction like few had seen before or after.

A Brief History of the Transistor

Something interesting happened as Henry Ford and other titans in the late nineteenth and early twentieth centuries demonstrated the power of scale economics. Innovation grew increasingly complicated and expensive, making it hard for an individual tinkerer to pursue it. Organizations began to hire groups of scientists. A new era of innovation, the era of the corporate lab, was upon us.

One of history's most famous corporate labs was Bell Telephone Laboratories, better known as Bell Labs. In the mid-1920s, a focused research effort at AT&T developed thermionic vacuum tubes that amplified electrical signals, enabling the first transcontinental call.* Inspired by this effort, AT&T decided to set up an industrial research center and staff it with thirty-five hundred people.

The transistor troika—Shockley, Bardeen, Brattain—each followed eclectic paths to Bell Labs.

---

* Kids, in ye olden times, making a telephone call required a physical connection between the two phones. A vast network of wires, switches, and amplifiers enabled this. One innovation when I was young was a special phone number that would ring twice when dialed. We had one number for outsiders and another number for family members. Both numbers rang the same landline. So when we called that second number, our parents would actually answer the phone. Times change.

Shockley could trace his lineage back to John Alden, the cooper on the *Mayflower*, the ship that brought the Pilgrims from England to the United States in 1620.* Born in London, Shockley moved to San Francisco as a young boy. When he was thirteen, his family moved to Los Angeles, and the frontier nature of the quickly growing city shaped his character. Some years later, he took a cross-country road trip to start graduate school at the Massachusetts Institute of Technology. His car of choice was a 1929 DeSoto roadster (no Model T for Shockley), and he drove with a loaded pistol. Hollywood to the core.

John Bardeen, a soft-spoken and cerebral scientist, was, like Henry Ford, a natural tinkerer. In the early 1920s, he started experimenting with radio, building his own set. He finished high school when he was fourteen. In 1933, he went to graduate school at Princeton University, starting in mathematics and then shifting to physics.

In *Crystal Fire*, a history of the transistor and the primary historical guide for this chapter, scientist-historians Michael Riordan and Lillian Hoddeson describe Walter Brattain as a "salty, silver-haired man who liked to tinker with equipment almost as much as he loved to gab."[1] Brattain was born in China, where his father taught at a private school for Chinese boys. He grew up in Spokane, Washington, on a cattle ranch and then attended Whitman College in Walla Walla, Washington. While the school lacked a cutting-edge physics department, one of the school's physics professors inspired and motivated Brattain to deeply explore the field.

In 1924, while browsing some literature in a tiny research laboratory, Brattain came across the *Bell System Technical Journal*, published by Western Electric, the manufacturing arm of AT&T. A series of articles titled "Some Contemporary Advances in Physics" summarized developments in Europe. "To a young man majoring in physics," Brat-

---

* Fun fact. So can I! Apparently. 1620. What else happened that year? Sir Francis Bacon's book. Cue eerie music.

tain later recalled, "these were very provocative articles about the newest developments."[2]

In 1938, Mervin Kelly, the head of the Bell Labs vacuum tube department, created a small team under Shockley to look at solid-state replacements for vacuum tubes. Vacuum tubes had clear benefits. They amplified power effectively, and they could be reliably manufactured. But they also had significant limitations. They burned out and needed to be replaced. It took time for them to heat up. And they gave off a lot of heat, running the risk of damaging other equipment. Solid-state technologies theoretically could be smaller, cooler, and more reliable.

At the end of 1939, Shockley scribbled on a piece of paper, "It has occurred to me that an amplifier using semi conductors rather than vacuum is in principle possible."[3]

Nonlinear Paths to a Breakthrough

But then . . . life happened. More specifically, World War II happened. By 1942, seven hundred people in Bell Labs were working on military projects. Eventually, 90 percent of Bell Labs' staff were focused on the military. Was the transformation to military production a distraction that delayed important commercial development? That's one way to frame it. My view is that it created a unique opportunity to push different technological frontiers, generating new knowledge that ultimately dramatically accelerated the development of semiconductors.

Shockley, for example, became the research director of the Anti-Submarine Warfare Operations Research Group in 1942. He focused on radar technology. Building effective radar systems requires detailed knowledge about the purification and manipulation of silicon and germanium. Those materials are known as semiconductors because they are perfectly imperfect conductors of electricity. What does that mean? Glass insulates, which means no electricity flows through it. Copper conducts, which means electricity flows through it unabated. A semiconductor does either, depending on the circumstance. That meant that a semiconductor could, theoretically, control an electrical current.

In 1945, when World War II ended, Bell Labs moved into a new facility in New Jersey. With office space scarce, researchers had to share small offices. Bardeen reported enjoying the seeming constraint, because it provided opportunities to look over the shoulders of experimenters and talk with them about their results. There were regular gatherings, or *chalk talks*, where the semiconductor group would kick around ideas that individual members were exploring. The team mixed theoretical, experimental, and technical expertise. Brattain clearly valued these sessions, noting how the diverse expertise encouraged free-flowing discussions that "went to the heart of many things" and led to immediate action.[4]

Bardeen and Brattain began experimenting with what ultimately was called the point-contact transistor. On December 16, 1947, Bardeen told his wife Jane, "We discovered something important today."[5] As he didn't talk about work at home very much, Jane knew it was important indeed. Shockley was overseeing the two scientists, sharing his own ideas about how to design a transistor so that it could be easily manufactured.

A week later, Bardeen and Brattain ran an experiment for Shockley's boss, acoustics experts Harvey Fletcher, and Bell Labs' research director Ralph Brown. Authors Riordan and Hoddeson describe how the proof-of-concept transistor was less than an inch high. It featured a small strip of gold foil cut by a razor and a "makeshift spring fashioned by a paper clip" and was "clamped clumsily together by a U-shaped piece of plastic resting upright on one of two arms."[6]

The signal on an oscilloscope jumped whenever Brattain toggled the contraption into a circuit. That meant the current was being amplified. It worked!

In a way, the experiment was a win for the team, for the department, and for Bell Labs. But in another, more accurate way, it was a win for Brattain and Bardeen.* Shockley argued that he had participated enough in the point-contact transistor to have his name included on the patent. Brattain recalled that Shockley said, "Sometimes the people who do the work don't get the credit for it." Brattain barked back, "Oh hell, Shockley. There's enough glory in this for everybody!"[7] Bell Labs patent attorneys decided that Shockley's contribution was at best marginal and dropped him from the patent application. Bardeen and Brattain had their names on a patent filed June 17, 1948, for a "three-electrode circuit element utilizing semiconducting material." The patent office awarded the duo patent 2,524,035 on October 3, 1950.

Patents matter for scientists, a lot. While Shockley certainly could feel proud that his research team invented a truly breakthrough technology, he was personally disappointed. "My elation with the group's success was tempered by not being one of the inventors," he said. "I experienced frustration that my personal efforts, started more than eight years before, had not resulted in a significant inventive contribution of my own."[8]

A month after the successful test of the point-contact transistor, Shockley scribbled in his notebook an idea for a "semi-conductor valve." One month after that, researcher John Shrive described a puzzling result for an experiment he had just run. Shockley realized that the result fit the idea in his notebook and announced it to the group. Further refinement led to Shockley's own patent application for a "junction" transistor, which promised to significantly simplify Bardeen and Brattain's point-contact transistors. That simplification was a critical step to take the technology from the lab to the market.

As the head of the research group, Shockley could direct further resources toward his own path of discovery. He pursued more patents

---

* This is a veiled *Simpsons* reference from "Deep Space Homer." It makes me miss my late brother, Mike, who could always find an appropriate Simpsons reference. After all, everything's coming up Millhouse!

and published technical descriptions of his work. He went from frustration to elation as his star rose.

Bell Labs officially announced the technology to the world on June 30, 1948. Here's how Brown introduced it:

> *What we have to show you today represents a fine example of teamwork, of brilliant individual contributions and of the value of basic research in an industrial framework.*
>
> *We have called it the Transistor, T-R-A-N-S-I-S-T-O-R, because it is a resistor or semiconductor device which can amplify electrical signals as they are transferred through it from input to output terminals. It is, if you will, the electrical equivalent of a vacuum tube amplifier. But there the similarity ceases. It has no vacuum, no filament, no glass tube. It is composed entirely of cold, solid substances.*[9]

The *New York Times* was so moved that it put the news on page 46 in a regular feature called "The News of Radio." Readers first learned about two new shows CBS was debuting over the summer: *Mr. Tutt* and *Our Miss Brooks* ("with Eve Arden playing the role of a school teacher who encounters a variety of adventures"). And then, a few paragraphs about the transistor.[10]

The transistor changed the world. But not overnight.

The new technology had unique benefits that existing technologies did not. It was small, it didn't emit heat, it was cheap, it started instantly, and it used energy incredibly efficiently.

The transistor also had unique limitations not found in existing technologies. It was sensitive to environmental conditions. It wasn't totally reliable. A 1949 *Consumer Reports* piece stressed how the devices were noisy and noted that Bell Systems' statements were "subdued." As Jack Morton, who led the development of solid-state devices at Bell Labs, said, "In the very early days, the performance of a transistor was apt to change if someone slammed a door."[11] And it couldn't match the raw power provided by vacuum tubes. Plus, it couldn't simply be dropped into existing products—those products would have to be completely redesigned around a transistor.

New benefits. New limitations. Classic disruptive innovation.

Companies that manufactured televisions, floor-standing radios, and networking equipment purchased a license for the transistor but proceeded cautiously. The technology looked inferior to existing solutions. Incorporating transistors would require substantial design work. The path from experiment to real-market application started in an unexpected place.

The Explosion of Hearing Aids

One of the enabling powers of disruption is the ability to open previously constrained markets. The would-be disrupter must pinpoint which market to start. The transistor, based on perfectly imperfect semiconducting material, shows the importance of finding a market segment that views the disruptive solution as . . . perfectly imperfect.

A disruptive innovation's imperfections are limitations that mainstream solutions don't have. In the right context, however, the innovation is perfect, because it has the potential to bring something uniquely powerful to a target customer that historically found existing solutions expensive, inconvenient, or sometimes simply inaccessible.

Consider the hearing aid. Prior to the transistor, the market was tiny. Using vacuum tubes to power the devices required that the user wear a bulky pack on the waist with a cord to the device in their ear. That pack would heat up. Vacuum tubes burned out and had to be replaced. All in all, vacuum-tube-powered hearing aids were inconvenient, expensive, and uncomfortable.

It isn't surprising, then, that hearing aid companies began using transistors as early as it was commercially possible. They were aided by a historical quirk. Alexander Graham Bell, the progenitor of the AT&T

system that housed Bell Labs, was passionate about helping people who were hard of hearing, and Bell Labs offered royalty-free licenses to hearing aid manufacturers. Hearing aid companies therefore had an easy way to play around with the technology.

In late 1952, Sonotone started selling $229.50 hearing aids ($2,700 today) with one of three vacuum tubes replaced with a transistor made by Germanium Products Corporation. Acousticon then introduced a tubeless single-transistor model for $74.50 ($870 today). Raytheon aggressively pursued this market, creating the CK718 transistor, the first mass-produced transistor. In a few short years, more than 95 percent of all hearing aids used transistors.

The transistor allowed for smaller, cheaper products. Its lower power consumption decreased annual battery costs from $100 to $10 ($1,100 to $110 today). The market for hearing aids expanded dramatically.

The small size of the hearing aid market meant the general public still had little idea of the transformational power of the transistor. Riordan and Hoddeson noted that most consumers still purchased millions of radios and TVs produced by "such grayflannel giants of American Industry as General Electric, Philco, RCA, and Zenith" with the transistor "largely perceived as an expensive laboratory curiosity with only a few specialized applications."[12]

"Grayflannel giants." At the time, the communications industry was dominated by companies with three-letter abbreviations: AT&T, IBM, RCA, and CBS. That all began to change in the early 1950s.

The Transistor Radio

In 1954, Pat Haggerty, a vice president at Texas Instruments (TI), advocated for the idea of a transistor-based, pocket-size portable radio. "I tried, without success, to interest several of the established large radio manufacturers in the idea," Haggerty recalled.[13] Haggerty's proposed product looked too expensive, and the transistor's performance was still just too inconsistent.

So, TI decided to launch its own radio. It partnered with Industrial Development & Engineering Associates (IDEA) to introduce a portable radio in time for Christmas 1954. It shipped fifteen hundred radios by the end of 1954 and topped a hundred thousand by late 1955. The radio's $49.95 price ($570 today) was attractive to consumers, but TI struggled to profit at those price points. It decided to shift its focus to providing components. The tipping point moment for the transistor radio came, surprisingly, from the Tokyo Tsushin Kogyo company.

Founded by Masaru Ibuka and Akio Morita after World War II, the company enjoyed its first successful market in tape recorders for courts and schools. Ibuka was initially skeptical about the transistor, but after visiting the United States to attend a seminar with Bell Labs, he changed his mind. By 1953, the company agreed to pay the $25,000 licensing fee—about $300,000 in modern terms and a stunning 10 percent of the company's net worth. The first product the company introduced in Japan was the TR-55 radio, which featured five in-house-developed transistors. The TR-63, a cheaper model that was small enough to fit in a pocket, was launched globally in 1957.

At the time, the world was fighting a pandemic, the so-called Asian flu that had originated in China in early 1957, spread to Singapore, and then appeared in the United States near the end of 1957. The virus went on to kill 160,000 people in the United States and an estimated 1 to 4 million globally. The pandemic coincided with a steep recession in the United States.

These events didn't stop the TR-63. The product had clear and obvious limitations compared with floor-standing radios. Audio quality, for example, could best be described as tinny. But the radio provided new benefits. It was affordable and portable. Teenagers could listen to baseball games or rock and roll out of the earshot of disapproving parents. It went on to sell more than seven million units. Seeing the device's stunning results, Toshiba and Sharp soon jumped into the market, helping to power Japan's economic miracle of the 1960s and launch the communications and information age.

Was Tokyo Tsushin Kogyo a historical footnote, overtaken by Toshiba and Sharp? No. Ibuka and Morita knew they needed a simpler name to expand the company globally. They started with the Latin word for "sound"—*sonus*. At the time, bright people were called *sonny*, but saying "sohn-nee" in Japanese means "to lose money." "We pondered this problem for a little while," Morita said, "and the answer struck us one day. Why not drop one of the letters."[14] And the brand Sony was born.

The army of American three-letter abbreviations was about to face an onslaught of new competitors from Japan and Korea, drastically changing industry dynamics.

Shockley's Next Act

Let's go back to our friend Shockley. It was 1953. Shockley was personally and professionally restless. He bought a green Jaguar XK120 super sports convertible and toured Europe. He dabbled at teaching. He tried to go back to a government research program. He expressed frustration with the limits he faced at Bell Labs, writing in a letter to his mother that "recognition of outstanding individual contribution is frequently lacking when organizational actions are taken."[15]

So, in 1955, he left Bell Labs to create a company to commercialize semiconductors. The first project was to mass-produce the transistors that he had helped create. Frederick Terman, the provost and dean of engineering at Stanford, helped the company set itself up in a part of the country that is now known as Silicon Valley. Shockley sought to attract the best PhDs in the United States, and his reputation certainly helped with that quest. "It was like picking up the phone and talking to God," recalled Robert Noyce, an eminent physicist and early hire who would go on to leave his own mark on the industry. "He was absolutely the most important person in semiconductor electronics."[16]

Noyce was joined by seven other young scientists: Gordon Moore, Julius Blank, Victor Grinich, Jean Hoerni, Gene Kleiner, Jay Last, and Sheldon Roberts. It turned out, however, that Shockley had what Moore called "some very peculiar ideas about motivating people."[17] He administered psychological tests, and he could be paranoid. One time, he subjected two technicians he suspected of wrongdoing to polygraph tests. As the company struggled to successfully manufacture semiconductors, he became increasingly sleep deprived, a condition that only fed his paranoid tendencies.

In September 1957, Shockley's erratic and paranoid behavior became too much. Noyce led the group of eight—which Shockley branded the Traitorous Eight—to jump ship. They officially resigned on September 18, 1957. Twenty-four hours later, with help from Arthur Rock from the New York bank Hayden, Stone & Co., the group received investment from Fairchild Camera and Instrument to create Fairchild Semiconductor.

Before describing how Fairchild helped spur the development of the personal computer and venture capital industries, let me offer a brief postscript on the three Nobel Prize winners.

After his work on the transistor, John Bardeen shifted his research focus to superconductivity. Although he left research to teach in 1951, the impact of his research continued. In 1972, he became the first physicist to win two Nobel Prizes. He served on the board of directors for Xerox Corporation in the 1960s and 1970s and was a member of scientific advisory committees in three separate presidential administrations. He died in 1991.

Walter Brattain stayed at Bell Labs until 1967. In the early 1960s, he also started teaching at Whitman College. Later in life, he claimed that his only regret about the transistor was "its use for rock and roll music."

I still have my rifle and sometimes when I hear that noise, I think I could shoot them all."[18] He died in 1987.

William Shockley never had another chance to commercially cash in on his invention. He taught for the next two decades. The dark side of the drive that leads disruptive innovators to great heights made its appearance when he wandered into discussions about race relations in the 1970s, opining that Black people were genetically inferior to white people. Shockley died in 1989. Riordan and Hoddesdon sum up his life: "In the process of turning silicon into gold many others became millionaires—and a few even billionaires. But due to fate and his own obstinacy, Shockley never got a chance to enter this Promised Land himself. That is why he truly deserves the title given him by his long-time friend and old traveling companion Fred Seitz: the Moses of Silicon Valley."[19]

Intel and the Rise of Venture Capital

A key part of the pattern of disruptive innovation emerges through the transistor's story. It is predictable that disruption will start in particular contexts, among customers who are overshot by existing solutions or who face a barrier to consumption. It is unpredictable, however, *which* specific contexts a disruption will begin. Serving markets that are currently *not* consuming can present a particular challenge. The markets are either small (hearing aids) or nonexistent (transistor radios). For this reason, they are difficult, arguably impossible, to measure and size. An organization that demands proof before making any investment decision simply can't back an idea that follows this path.

The transistor helped accelerate the development of a form of investment—venture capital—to enable a new generation of innovators to realize those opportunities.

In the era of the corporate lab, large organizations demonstrated the patience to handle the twists and turns that characterized the path to innovation success. However, two parallel developments led to fundamental changes in the source of financing for innovation.

The first development was the ossification of the large organization. Modern management techniques had many benefits, but the quest to routinize and stamp out variation ran counter to the controlled chaos that spurs disruption. The postwar boom created ample opportunities for big organizations to globalize and become enormous. In doing so, however, large organizations became increasingly hostile to innovation.

The second development related to the rise of individualism and the counterculture. Rock and roll (listened to on those transistor radios) rose in popularity. Hair length increased. People questioned authority. People sought new forms of freedom and expression. Call it disruptive innovation of a social variety.

The blossoming of the modern semiconductor industry and its first global champion, Intel, parallels these developments and ushers in a new era of innovation, the era of the venture capital–backed startup.

Fairchild Semiconductor had enjoyed significant successes during the 1960s. However, in the late 1960s, Moore and Noyce again became unsettled, feeling overly shackled by Fairchild's setup. The two again ejected themselves, along with a young engineer named Andy Grove, and decided to set up a new company. Rather than seek a corporate sponsor, they wanted to create their own company. Now, they faced a quandary. They wanted the freedom of their own company but needed substantial funding to get their business off the ground. What to do?

A new asset class was emerging. For a few decades, wealthy individuals had been experimenting with vehicles to invest in risky young companies. Then, in 1968, a year before the music world had Woodstock and the technological world had the moon landing, American Research and Development Corporation (ARD) showed the power of early-stage investing.

A number of Boston business and intellectual leaders founded ARD in 1946. In 1957, it invested $70,000 in the Digital Equipment Corporation (DEC), a startup seeking to develop smaller, cheaper alternatives to mainframe computers sold by IBM (yes, a disruptive innovation!). That stake was worth more than $350 million when DEC went public in 1968. That's a compound annual return of 117 percent!

So, when Moore and Noyce were looking for funding, they had choices like Draper, Gaither & Anderson, Silicon Valley's first limited liability partnership, formed in 1959; Greylock Partners, formed in 1965; or their old friend Arthur Rock, who had shifted from banking to venturing activity. And Rock is whom they went with. He invested $10,000 of his own money, became Intel's chair, and within forty-eight hours raised $2.5 million in loans that would convert to stock in the case of an initial public offering (IPO), which Intel did . . . two years later.

One year after Intel's IPO, Kleiner Perkins Caufield & Byers and Sequoia Capital were founded (and yes, it is the same Kleiner who was part of the Traitorous Eight). These and similar institutions helped support the formation of Apple, Genentech, Microsoft, Cisco Systems, Amazon, Facebook, and Google—all companies that rode to some degree on the invention by Bardeen, Shockley, and Brattain.

The Transistor's Echo: M-PESA and Competing against Nonconsumption

Apple cofounder Steve Jobs said, "Customers don't know what they want until we've shown them."[20] It's true. Customers will say all sorts of things in research settings. They'll say they want things that they don't, claim they will do things that they won't.*

---

* I won't repeat it here, but there's a fun story in my book *The First Mile* about a failed idea Innosight had to create a medical tourism venture. The story ends with the punch line "people lie."

That doesn't mean that innovation is a guessing game. People have problems they are trying to solve, what Clay and other researchers later called "jobs to be done." Products and services that make it simpler and easier for people to get important jobs done thrive. Those that make people's lives more difficult or that require them to fundamentally change their preferences or their behavior often struggle.

The transistor found paths to a market where its limitations were irrelevant and its benefits were boons. It brought hearing aids to a wider population and enabled the dramatic miniaturization of consumer electronic devices. And ultimately replaced the vacuum tubes in networks. Keep this journey in mind the next time you see news of a hot new technological breakthrough. The path to success is rarely linear, and many technologies follow the transistor's path of starting in unanticipated spaces.

Clay ultimately dubbed this approach "competing against nonconsumption," when a disruptive innovation's simplicity and affordability enable more people to consume something in more places. Disruptive success comes from overcoming common barriers to consumption, most notably wealth, skills, and access.

A modern example—the development of M-PESA, a mobile-phone-based money transfer service in Kenya—shows the power of this approach. In the aughts, most Kenyans lacked access to a bank account. Transferring money was complicated and expensive. The best that most customers could do was to use alternatives like *hawala*, a network of brokers operating on the honor system.

In 2003, Nick Hughes, the head of social enterprise at Vodafone, earned a £1 million grant from the UK government for a proposal to use mobile phones to deliver financial services. He organized workshops in Nairobi, Kenya, and Dar es Salaam, Tanzania, for people interested in new banking solutions. Susie Lonie, who had a background in mobile commerce, joined the effort to oversee on-the-ground efforts to commercialize Hughes's concept.

"The difficulty with doing something totally new is that you are not terribly sure where you are going until you arrive," Hughes and Lonie wrote in a 2007 summary of the launch of M-PESA. "We knew a few things, but much of the time we were guessing."[21]

The two worked with Safaricom, a Kenyan mobile phone operator partly owned by UK-based Vodafone. They spent time in the market to understand the myriad work-arounds Kenyans followed to handle transactions. Hughes and Lonie developed a simple, affordable solution. Customers would register with an agent in accessible locations like Safaricom stores, food markets, or gas stations. Registered users could then transfer money using their mobile devices to anyone with a mobile device, even if the other user was not a Safaricom subscriber. M-PESA (*M* for mobile and *PESA*, the Swahili word for money) users paid low fees for the service, and Safaricom pooled deposits and placed them in regulated banks, protecting the assets.

Three years after its 2007 launch, 40 percent of Kenya's adult population was using the service, and close to twenty thousand retail outlets accepted M-PESA payments. Safaricom expanded into bill payments, paychecks, and international transfers.[22] Over the next two decades, similar disruptive developments created new classes of financial services providers in many other emerging markets.

Anytime there is a barrier to consumption, where consumption requires wealth, specialized skills, a centralized location, or significant amounts of time, there are opportunities for simple, convenient, disruptive solutions.*

It can take a while for a disruptive innovation to have its full impact. Such innovations often look innocent in their early days, showing up

---

* One design decision was to not make this a tool book. My earlier book, *The Innovator's Guide to Growth*, presents several practical techniques to identify nonconsumption. It features the acronym SWAT, which refers to barriers related to skills, wealth, access, and time. An alternative is WASTE, which adds "existing behaviors." I don't like the latter term, because although an existing behavior demonstrates that a problem is important, it doesn't feel like the same kind of barrier. Your call.

in novel use cases like hearing aids and pocket-size radios. Never discount the power of democratization. Innovations that make the complicated simple and the expensive affordable can go on to change the world, even if their first public mention is buried beneath banal media updates on page 46 of the *New York Times*.

7

The Recipe for Disruption

Julia Child and the Art of French Cooking

It started out with a kiss.

Wait a second. That's from the Killers' song "Mr. Brightside," which somehow went from being a fine rock song when it was released in 2004 to a cathartic dance-floor-filling anthem two decades later.

It started out with a fish.

Yes, that's what I was looking for. That's where the Julia Child story starts. With a fish. With sole meunière, to be specific. A key part of what I call "the meal that changed everything" (figure 7-1).

It was November 3, 1948. Julia and her husband Paul had just arrived in France. The pair had met while working at the Office of Strategic Services (OSS), a precursor to the Central Intelligence Agency. Yes, Julia Child had a turn as a spook. Kind of. She was in back-office roles

Figure 7-1

Menu from "the meal that changed everything," November 3, 1948

Menu from La Couronne: *that lunch* in Rouen, November 3, 1948

Source: Reprinted with permission from Bob Spitz, *Dearie: The Remarkable Life of Julia Child* (New York: Vintage, 2013).

in modern-day Sri Lanka and China but had access to a wide range of confidential information. Paul, a poet, photographer, and longtime mid-level State Department employee, was assigned to run the Visual Presentation Department for the US Information Service in Paris. He had previously lived in France and spoke the language fluently; these two accomplishments would aid him in his quest to promote French–American relations through the visual arts.

Consulting the Michelin Guide, the couple arrived in la Place du Vieux Marché in Rouen. It was the square where Joan of Arc had been burned at the stake. The Michelin Guide gave La Couronne (the Crown), a restaurant in a house built in 1345, three forks and a spoon.

"It was warm inside," Julia recalled, "and the dining room was a comfortably old-fashioned brown-and-white space, neither humble nor luxurious. At the far end was an enormous fireplace with a rotary spit, on which something was cooking that sent out heavenly aromas."[1]

Julia had a glass of wine, a new lunchtime experience for her. The air was filled with the intoxicating aromas of shallots, an unnamed sauce being reduced on the stove, and the acidic smell of salad being tossed with lemon, olive oil, wine vinegar, salt, and pepper in a bowl.

The oysters had a "sensational briny flavor and a smooth texture that was entirely new and surprising." They were accompanied by *pain de seigle*, a pale rye bread with luxurious butter. And then, that fish:

> *It arrived whole: a large, flat Dover sole that was perfectly browned in a sputtering butter sauce with a sprinkling of chopped parsley on top. The waiter carefully placed the platter in front of us, stepped back, and said: "Bon appétit!"*
>
> *I closed my eyes and inhaled the rising perfume. Then I lifted a forkful of fish to my mouth, took a bite, and chewed slowly. The flesh of the sole was delicate, with a light but distinct taste of the ocean that blended marvelously with the browned butter. I chewed slowly and swallowed. It was a morsel of perfection.*[2]

A perfect meal, indeed, the most exciting meal in Julia's thirty-six-year life.

It certainly wasn't clear on that Wednesday that Julia would go on to transform the world of cooking. That her meticulous, scientific approach would follow the classic disruptive pattern of enabling a broader population to do something that was once reserved for experts. That she would write more than a dozen bestselling books. That she would become arguably the world's first reality television star. All that was clear on that day was that the fish was delicious.

Bon appétit!

A Brief History of Julia Child

It's doubtful many people would have predicted Julia Child's rise as a culinary celebrity who would be beamed into millions of households. Born in Pasadena, California, she was a towering six feet, two inches tall (the average woman at the time was about five feet four). She had a voice described by biographer Bob Spitz as "tortured and asthmatic, with an undulating lyrical register that spanned two octaves."[3] It was "a high-pitched warble. . . . The family called it 'hooting,' the result of unusually long vocal cords, which gave the voice a kind of comical slide-whistle effect."[4]

And perhaps most notable of all, she was a disaster in the kitchen.

One childhood friend described how an effort to create strawberry jam led to a "gooey mess . . . though, God knows, she tried."[5] When Julia met the family of Paul's twin brother Charlie, Charlie's son Jon remembered: "Word around our house was Paul's girlfriend couldn't cook. The joke was: she could burn water if she boiled it."[6] Julia took a bride-to-be cooking course from two Englishwomen in Los Angeles, learning about basic dishes like pancakes. The first meal she decided to cook for Paul was anything but simple: brains simmered in red wine. In her own words, "The results, alas, were messy to look at and not very good to eat. In fact, the dinner was a disaster."[7]

The move to France and that fateful meal sparked something.

Julia Child's Hero's Journey

There's a pattern connecting the disruptions we have studied to date. Florence Nightingale isn't allowed to go to nursing school. Henry Ford gets kicked out of a company that bore his name. William Shockley doesn't make it onto the first patent submission. Our hero starts a quest, suffers a setback, and, with the support of a mentor, helper, or friend, emerges triumphant. And indeed, every disruption I've ever studied fits the *monomyth*, the basic pattern that connects religious texts and historical legends. As described in *The Hero with a Thousand Faces* by Joseph Campbell, the specifics are unpredictable, but the pattern is predictable. Julia Child's story fits the hero's journey archetype beautifully.

It starts when the hero gets a call to adventure.

Child's call came in 1951. She was coming off a shocking failure. Earlier that year, she had failed the final exam at Le Cordon Bleu, the legendary French cooking school. At the school, she had studied under Max Bugnard, a septuagenarian who focused on making recipes simple and bringing joy to the kitchen. Julia remembered him saying, "Yes, Madame Scheeld ["Child" pronounced with a French accent], fun! Joy! You never forget a beautiful thing that you have made. Even after you eat it, it stays with you—*always*."[8] What happened on the final exam? Julia was asked to cook three dishes that had not been part of her program but were in Le Cordon Bleu's cookbook. She didn't know the recipes and butchered them. Never matter. She memorized the basic cookbook, retook the test, aced it, and received a diploma that was backdated to the date of the first test.

Later that year, Child was introduced to Simone Beck Fischbacher, otherwise known as Simca. A few days later, she met Louisette Bertholle. Both were members of Le Cercle des Gourmettes, an organization founded in 1927 to encourage female chefs. The three would go on to create a program called l'École des Trois Gourmandes, where they taught classes for women looking to improve their cooking.

In late 1951, Fischbacher and Bertholle told Child that they had been working on a cookbook of French recipes targeting the US market. A US publisher named Ives Washburn had slapped together *What's Cooking in France*, a booklet based on the recipes. The $1.25 book sold about two thousand copies. Aspiring to do more, Fischbacher and Bertholle drafted a more extensive book that mixed recipes with descriptions of the practice of a French chef.

They shared the book with Dorothy Canfield Fisher, an author and a member of the editorial board of the Book of the Month Club.* She panned it. "Get an American who is crazy about French cooking to collaborate with you," Fisher wrote in a letter," somebody who both knows French food and can still see and explain things with an American viewpoint in mind."[9]

Who could that be?

Child considered the idea. She reflected on what she saw as an opportunity. Most cookbooks were inexact, leaving the details open to the interpretation of the reader. That approach led to inconsistent results, even in the hands of talented chefs. And chefs who lacked talent, well, forget about it. She saw how Fischbacher had a set of wonderful recipes and was a precise, thoughtful chef, and how Bertholle's instincts could lead to unexpected flourishes—shallots here, fresh peas there—and impressive results. When, in August 1952, Ives Washburn told Fischbacher and Bertholle that the editor who was working on the book had quit, the two asked Child if she would join the writing team. She immediately agreed.

The goal was to create a serious book but one that made cooking approachable and embraced Max Bugnard's joy. One that broke complicated French cooking down to its logical steps, enabling any reader to both understand French cooking and execute it well. No cookbook had ever done this, in English or in French. Ives Washburn asked for a polished manuscript by March 1, 1953.

The next stage of the hero's journey involves an unexpected enabler, when the hero receives surprising support from a mentor or helper. Child's was Avis DeVoto.

---

* What's the Book of the Month Club, you ask? Founded in 1926, it was a subscription service for avid readers. Members would receive a curated number of books in the mail monthly, along with a newsletter describing other recommended books (that they could purchase, of course). Book of the Month was rebooted as an e-commerce service in 2015.

In the spring of 1952, Child had read an article complaining about the state of American-made cutlery in *Harper's Magazine*. She found the article so moving, she wrote a letter to its author, Bernard DeVoto. She got a response not from Bernard but from his wife, Avis. The two women became frequent correspondents and lifelong friends.

The DeVotos were well connected to the publishing world. Child sent Avis a draft chapter on sauces, and Avis asked about showing it to her husband's publisher, Houghton Mifflin, which happened to have on its staff a cooking expert named Dorothy de Santillana. Ives Washburn had been sitting on a draft of the manuscript without commenting on it for weeks, so Child and her team agreed to submit the manuscript to this other publisher. In January 1953, Houghton Mifflin offered an advance of $750 ($9,000 in modern terms) against a royalty of 10 percent. Houghton Mifflin hoped to publish in June 1954. At the time, Child thought the earliest the book would be published was 1955.

She was off by six years. It was time for the third stage of the hero's journey: the darkness of an abyss.

Most of the work on the cookbook fell on Child's and Fischbacher's shoulders. The two iterated and experimented, sharpening and perfecting recipes. The tentative title was *French Cooking for the American Kitchen*. It expanded to seven hundred pages. Julia and Paul Child moved back to the United States in 1956. Their move made collaboration with Fischbacher more challenging but allowed Julia to see the reality of the world her target reader inhabited. US veal was less tender than French veal. Its turkeys were larger. And it seemed that Americans ate more broccoli than the French did.*

In 1958, Houghton Mifflin wanted to sit down and review the working draft of the manuscript. Fischbacher came over to the United States. She and Child took an eleven-hour bus ride from Union Station

---

* Broccoli. My nemesis as a child. I refused to eat dinner if broccoli was involved. My eight-year-old son, Teddy, loves it. Go figure.

in Washington, DC, to Boston through a blinding snowstorm to make a February 24 meeting with Houghton Mifflin. They took turns sleeping while protecting their precious manuscript. In the back of their minds may have been a voice of doubt. They had been shopping recipes from the book with American magazines and received no interest. "The editors seemed to consider the French preoccupation with detail a waste of time, if not a form of insanity," Child said.[10]

Perhaps they shouldn't have been that worried about protecting the manuscript. De Santillana politely but firmly slammed the book, writing, "This is not the book we contracted for, which was to be a single volume book which would tell the American housewife how to cook in French."[11]

Child had only herself to blame. The project had gotten away from her. She had provided seven hundred pages that *only* covered sauces and poultry. A full cookbook covering eggs, fish, desserts, vegetables, and more promised to be more than fifteen hundred pages. That cookbook would be unprintable, unusable, untenable.

She wrote, and then tore up, a letter proposing canceling the contract. Instead, she wrote a second letter saying that they would do what Houghton Mifflin wanted: a shorter, snappier book. Over the next eighteen months they thoroughly rewrote the book. They expanded from sauces and poultry to a comprehensive primer on French cooking for serious cooks, covering everything from crudités to dessert. The length increased, but only by fifty pages.

They submitted the manuscript on September 1, 1959, Julia and Paul's thirteenth wedding anniversary. De Santillana wrote a positive note at the end of the month, saying she was "bowled over" at the intensity and detail in the book. Child had won over the editorial assistant. She was, however, about to face an even deeper abyss.

On November 6, 1959, editor in chief Paul Brooks threw a curveball. His note started positively, describing the book as "culinary art." However, he wrote that the book would be expensive to produce and that the team was skeptical there was a market for it, as the typical housewife might be "frightened" by it. "I am aware that this reaction will be a disappointment," Brooks wrote. "We will always be interested in seeing a smaller, simpler version. Believe me, I know how much work has gone into this manuscript. I send you my best wishes for its success elsewhere."[12]

Strike two.

In the final stage of the hero's journey, the hero emerges from the abyss transformed, triumphant.

Child reported quiet resignation at Brooks's note. "I wasn't feeling sorry for myself," she noted. "Even if we were never able to publish our book, I had discovered my raison d'être in life, and would continue my self-training and teaching."[13] But hope was not lost. Brooks suggested that Doubleday, a renowned publishing house, might be interested in the book. It isn't clear if he offered to make any introductions, but in any case, Avis DeVoto, reliable Avis, had a plan. Without asking Child, she sent the manuscript to Judy Jones, a young editor at Alfred Knopf publishing house.

When Jones lived in France, she was an editorial assistant for Doubleday. One day, a book with a picture of a young girl on the cover caught her eye. Her boss was planning to reject the book, but Jones couldn't put it down. She urged Doubleday to buy the rights for the book. The publisher did, and *Anne Frank: The Diary of a Young Girl* went on to sell more than thirty million copies and to be translated into seventy languages. When Jones returned to New York, she joined Alfred Knopf to work on translations of books from France.

"*French Recipes for American Cooks* is a terrible title," Jones wrote in a letter to her husband. "But the book itself is revolutionary. It could become a classic."[14]

What did Jones like about it? "She enjoyed our informal but informative writing style, and our deep research on esoterica, like how to avoid mistakes in a hollandaise sauce," Child wrote in her memoir.

"She congratulated us on some of our innovations, such as our notes on how much of a recipe one could prepare ahead of time, and our listing of ingredients down a column on the left of the page, with their text calling for their use on the right."[15]

Yes, that's right. Pick a cookbook off your shelf and flip it open. You are likely to see the ingredients in one column and the instructions in another. It just makes sense. But as always, it takes an innovator trying something different to come up with something that in hindsight looks blindingly obvious.* That innovator was Julia Child.

Alfred Knopf pounced. In mid-1960, Child, now stationed in Oslo, Norway, received a letter offering a $1,500 advance ($20,000 today) against royalties of 17 percent for the first twenty thousand sales, and 23 percent for sales beyond twenty thousand. After significant back-and-forth, they landed on the title *Mastering the Art of French Cooking*.

In June 1961, the young, telegenic president, John F. Kennedy, announced that René Verdon, a French chef, would serve as the White House chef. On October 16, Alfred Knopf released *Mastering the Art of French Cooking*. On October 18, page 47 of the *New York Times* featured a glowing review from influential restaurant critic Craig Claiborne with the article titled: "Cookbook Review: Glorious Recipes."**

> *What is probably the most comprehensive, laudable, and monumental work on [French cooking] was published this week. . . . It will probably remain as the definitive work for nonprofessionals. . . .*

---

* I teach my kids this principle by talking about wheels on luggage. It's hard to believe that it took until 1970 for someone (in this case, Bernard D. Sadow, the vice president of a Boston-area luggage company) to do something that now seems amazingly obvious.

** For those scoring at home, that is one page after the page announcing the invention of the transistor thirteen years earlier.

> *It is written in the simplest terms possible and without compromise or condescension.*
>
> *The recipes are glorious, whether they are for a simple egg in aspic or for a fish soufflé. At a glance it is conservatively estimated that there are a thousand or more recipes in the book. All are painstakingly edited and written as if each were a masterpiece, and most of them are.*[16]

An overnight success. Ten years in the making.

"In the same way that the space race tapped into Americans' desire to achieve something glorious in the realm of science, Julia tapped into a housewife's desire to expand the boundaries of her own world," biographer Spitz wrote. "Nobody knew American women were out there hungering for this, but out there they were. And Julia offered them an outlet for that pent-up ambition."[17]

"Save the Liver"

To support the launch of *Mastering the Art of French Cooking*, NBC's popular morning show *Today* asked for a cooking demonstration. Child and Fischbacher decided to cook an omelet. The only equipment needed was an electric hot plate. They heated their pan on the hot plate before the segment so that it was burning hot, and they nailed the live demonstration. This appearance began a remarkable three-decade run where Child and her warbly voice became a white-hot television star, further bringing cooking to wider audiences.

After the NBC demonstration, Child was asked to go on WGBH, a local public television station in the Boston area, where Albert Duhamel hosted the half-hour show *I've Been Reading*. The show received twenty-seven letters praising Child's appearance. In light of that audience response, WGBH asked twenty-eight-year-old director Russell Morash to do three half-hour pilot cooking shows. Paul and Julia practiced endlessly to prepare.

"So that we could rehearse," Julia Child said, "Paul made a layout sketch of the freestanding stove and work counter there, which we brought home and roughly emulated in our kitchen. We broke our recipes down into logical sequences, and I practiced making each dish as if I were on TV. We took notes as we went, reminders about what I should be saying and doing and where my equipment would be: 'simmering water in large alum. Pan. Upper R. burner'; 'wet sponge left top drawer.'"[18]

The trial shows clicked. The three pilots led to a twenty-six-show season. Maybe it was the preparation. Maybe it was the acting Julia Child did while a college student. Perhaps it was the teaching she did in France. Who knows for sure, but she went on to do ten seasons of *The French Chef* and several other shows. In 1966, she won an Emmy and was featured on the cover of *Time* magazine. She earned an advance of $25,000 ($250,000 in today's dollars) for a cookbook based on *The French Chef*. The book came out in 1968. The print run of one hundred thousand copies sold out on release. *Mastering the Art of French Cooking, Volume 2*, came out two years later. The first print run of a hundred thousand copies again sold out on release. Alfred Knopf was ready and had already set up a second run of fifty thousand copies.

What made Child work so well on television? Humor shines a light on that question. December 9, 1978, marked the eighth episode of the fourth season of *Saturday Night Live*, a show mixing short comedic sketches and live music performances. The first sketch, written by Al Franken and Tom Davis, featured Dan Aykroyd with a brown wig, glasses hanging around his neck, and a dish cloth at his belt, playing the part of Julia Child on *The French Chef*. He takes the liver out of the chicken in front of him, talks about the importance of saving that liver. He then accidentally cuts himself.*

---

* The sketch was based on a real incident when Child cut herself with a borrowed knife while cooking with Jacques Pépin. Could it possibly have been the same poor craftsmanship that Avis DeVoto's husband complained about in the early 1950s?

Blood spurts everywhere. Aykroyd talks about how to use a chicken bone to cauterize the wound. He attempts to use the phone in the back of the kitchen to call emergency help but announces it is a prop phone. As he collapses on the chicken, he gives Child's trademark send-off, the same words she heard at the meal that changed everything: "Bon appétit!" He rises and, one last time, says, "Save the liver."

Child loved the sketch. It captured in parody what made her so remarkable on television. Relatable. Approachable. Human. Yes, going through the technical details of a recipe but rolling with things when they didn't work out. One time, for example, a potato pancake fell on the worktable. Child scooped it up and put it back in the pan. "Remember," she said, staring directly at the camera in a way that made it feel like she was talking directly to the audience, "you're all alone in the kitchen and no one can see you."[19]

She reflected that she wanted her show to reflect reality. "Once I got going, I didn't like to stop and lose the sense of drama and excitement of a live performance," she noted in her memoir. "Besides, our viewers would learn far more if we let things happen as they tend to do in life—with the chocolate mousse refusing to unstick from its mold, or the apple charlotte collapsing. One of the secrets, and pleasures, of cooking is to learn to correct something if it goes awry; and one of the lessons is to grin and bear it if it cannot be fixed."[20]

Viewers loved her. "You are the only person I have ever seen who takes a realistic approach to cooking," one wrote. "By the time she gets to the table with her dish and takes off her apron," wrote another, "we are so much 'with' her that we feel as if someone had snatched our plates from in front of us when the program ends."[21]

Child would go on to author or co-author more than a dozen books in the next three decades. Typically, the books would intertwine with her television shows. It was a powerful, reinforcing model. Her last cookbook, *Julia's Kitchen Wisdom*, came out in 2000, when she was eighty-eight. She collaborated with David Nussbaum, the CEO of

America's Test Kitchen, which followed in Child's footsteps by following scientific principles to optimize recipes.

Julia Child died on August 13, 2004, two days before her ninety-third birthday. You can visit her kitchen in the National Museum of American History. Maybe you'll hear her distinctive voice. Maybe you'll smell the shallots she smelled in 1948. Maybe you will think of Child's basic philosophy: anyone can be a great cook.

"French cooking starts out from just perfectly direct principles," Child told *Time* magazine in 1966. "It's so important that there are reasons for doing things. It is a tradition with rules—perfectly simple ones. If you know them, then you can do any kind of cooking."[22]

Julia Child's Echo: The Disruption of Media

In May 2024, an unexpected thing happened in the Anthony household. We had a social media star in our midst. Our oldest son, Charlie, was a senior at Milton Academy. Every student at the school does a team-based project at the end of their senior year. Charlie's project involved exploring the stories of immigrants to Boston . . . by going with a friend to eat at a range of restaurants and interview the owners. Yes, it's that kind of senior project.

A group of five of Charlie's friends had a wacky idea: buy a school bus, renovate it into a bus/camper-van hybrid, and do a road trip across the United States. Charlie started spending weekends helping with the project. The group posted short clips of their work on Instagram and TikTok. One weekend, the number of viewers started to skyrocket. Why? Who knows? The algorithms liked the "Boys with the Bus." Tens of thousands became hundreds of thousands, became two million. The boys appeared on *Good Morning America*, a show with close to three million daily viewers, twice. Brands eager to reach their audience started calling. I was a bit investor in AssetCo (which owned the bus); the boys fully owned MediaCo (where sponsorship dollars went).*

All of this happened while I was researching and writing about Child, and I saw some beautiful symmetry. After all, Child democratized French cooking through do-it-yourself cookbooks and short instructional videos. The wave of disruption that has reshaped the media industry clearly fits this pattern.

I remember distinctly doing consulting work with Turner Broadcasting in 2006. At the time, the conglomerate had a range of popular television stations, such as TBS, TNT, CNN, and Cartoon Network. On October 9 of that year, Google paid $1.65 billion in stock for a fledging video-sharing site called YouTube. Turner executives scoffed at the seemingly lofty price. "If you add up all of the videos ever watched in YouTube's history, it equates to the eyeballs we get on a random Tuesday night," one executive told me.

At the same time, I was advising a number of newspaper companies. There were plenty of signals that the industry was thriving. I walked the hallways of gleaming headquarters. I attended off-sites in beautiful resorts. I might have even gotten a ride or two on a private plane. At the time, the newspaper industry was, believe it or not, wildly profitable.

What about Google and its quickly growing advertising business? "I'm not worried," an executive told me. "We have something they don't. We have feet on the street. We can go and work with the local businesses and help with their advertising strategy." I told them that Innosight (the consulting company at which I worked at the time) had purchased advertising from Google for years, paying for keywords like *disruptive innovation* and *innovation consulting*. We had never talked to a customer service representative, and we viewed Google advertising as a *great* service because it was simple and easy for us to use to manage our own campaigns.

---

* If you are wondering, yes, I got my investment back. Nice work, boys!

You know what happened next. YouTube, Instagram, and TikTok have exploded in popularity as it has become progressively easier for laypeople to create compelling content. More than 50 percent of young people surveyed in 2025 said they aspired to be social media "influencers."[23] Google's advertising sales grew exponentially from 2006. Newspaper companies suffered sharp financial losses and conducted mass layoffs. The industry in 2025 was a shell of what it was in 2006.

That's the power of disruption. When you make it easy for people to do things themselves, you create the possibility for explosive growth. That's good for those who enable the growth. It can be bad for people who prospered in an era of constraints.

And the Boys with the Bus? Charlie left the bus tour in South Dakota in July 2024 and flew home. He enjoyed his time in the spotlight but had other fish to fry. Thankfully, on our shelf there's an easy-to-use cookbook that can help with that.

Five Disruptive Ingredients

Julia Child's story offers the opportunity to examine an important pattern in disruptive innovation: disrupters' specific behaviors that drive success.

In 2011, Clay coauthored a book with Jeff Dyer from Brigham Young University and Hal Gregersen, who was then at INSEAD before moving to MIT in 2014. *The Innovator's DNA* detailed the results of a six-year research project by Dyer and Gregersen that involved a close study of the habits of twenty-five innovative entrepreneurs and surveys of more than three thousand executives and

---

* I always disliked the title, as DNA implies something that is deterministic, while the research clearly shows that most of what makes innovators successful is specific behaviors that can be learned. Studies on twins, for example, show that roughly one-third of creativity is from genetics. Yes, that means being otherworldly requires the right DNA. But it also means that just about anyone can get to a level of competency.

five hundred entrepreneurs and inventors.* When I was at Innosight, we combined *The Innovator's DNA* research with our own fieldwork to propose five behaviors that drive innovation success:

1. *Curiosity.* Great innovators question the status quo, looking for different and better ways to do things. They ask questions like "Why do we do things this way?" and "How might we do it differently?"

2. *Customer obsession.* You can't do something different that creates value unless you solve a problem that matters to a customer. Great innovators understand the people for whom they are trying to innovate, sometimes better than those people understand themselves. Innovators know their customers' hopes, dreams, expectations, and frustrations.

3. *Collaboration.* Great innovators recognize a line repeated from the printing press story: magic happens at intersections, where different mindsets and behaviors collide. They plant themselves at those intersections, recognizing that none of us is as smart as all of us.

4. *Adeptness in ambiguity.* The path to innovation success is never a straight line. There are positive and negative surprises along the way. There are false steps, fumbles, and, of course, failures. Innovation success doesn't come from planning and acting; it comes from testing, learning, and adjusting. Great innovators relish this journey and adeptly manage it.

5. *Empowerment.* You can't do something that creates value unless you actually *do something*. Great innovators think, ponder, pontificate, and plan. But more important, they *do*. They do everything they can to make their idea as tangible as possible, as quickly as possible. They ask for forgiveness, not permission.

The best illustration of these five qualities comes from what Child called "the Great French Bread Experiment."

Not surprisingly, the runaway success of *Mastering the Art of French Cooking* led to requests for a sequel. Fischbacher and Child had shelved so much from their original book, they certainly didn't lack raw materials.* Judy Jones from Alfred Knopf pushed them to include a recipe for a baguette.

Easy enough. Except for one pretty fundamental challenge. Boulangeries used special-purpose ovens for baking. You weren't going to get a special-purpose bread oven in an American kitchen, so what to do?

Thus, the Great French Bread Experiment, which Julia called "one of the most difficult, elaborate, frustrating, and satisfying challenges I have ever undertaken."[24]

Julia's *curiosity* led to her devouring two French textbooks on baking, going deep into how yeasts and flour worked. She *collaborated* by networking with experts. She read an article about Professor Raymond Calvel, an eminent baker and teacher. She contacted him and spent a day in the kitchen with him. What she learned blew her mind. His dough was "soft and sticky," and he let it rise twice in a cool place to three times its original volume to develop its flavor and texture. He had a specific way of folding and shaping the loaves, and he used a straight razor to slash the top of the loaves before putting them in the oven. The slashing opened up "the bread's gluten cloak and allowed a decorative bulge of dough to swell through the crust."[25]

How, however, to recreate the hot surface and steam that led to the baguette's unique look, feel, and taste? After all, her constant *customer obsession* meant it had to be something a normal person could do in a normal kitchen. Paul and Julia experimented, and experimented, and

---

* What happened to Louisette Bertholle? Child grew frustrated that she wasn't carrying her weight and essentially kicked her off the team. The Julia Child story is so fascinating. At one point this chapter was nearly nine thousand words long. If you want more stories about team dynamics, fish, sauces, early days in Paris, and more . . . you know where to find me.

experimented again. Thirty-one separate experiments in the end. A plastic spray bottle to squirt the loaves. Ice cubes in the bottom of the oven. The "wick system" with a wet bath towel extending out of a pan of water. "We even researched the medieval method of dampening a bundle of straw and throwing it into the oven to keep the air moist," Paul recalled.[26] *Adeptness in ambiguity* in action!

Sliding a tile made of asbestos cement on the oven rack created an affordable, effective baking surface. And dropping a hot brick, stone, or metal ax-head (if you happened to have one lying around) into a pan of cold water in the bottom of the oven created a beautifully effective puff of steam.

They had done it. And the recipe Julia submitted to her editor was only... thirty-two pages long. And used a material that scientists would later discover could cause cancer. Jones heard about that research two years after the first edition of *Mastering the Art of French Cooking, Volume 2*, came out. Child scrambled to find a replacement for the asbestos tile. On February 5, 1971, she experimented with three tiles: quarry tile, tortoise-glaze tile, and firebrick splits. All worked. Crisis averted.

In the end, the Great French Bread Experiment took two years, and according to Julia Child, used 284 pounds of flour. All the way through, Child demonstrated the *empowerment* of great disrupters. "No one is *born* a great cook," she wrote in the closing chapter of her memoir. "One learns by *doing*. This is my invariable advice to people: Learn how to cook—try new recipes, learn from your mistakes, and above all have fun!"[27]

Julia Child. A perfect innovator.

8

The Special Sauce

McDonald's Fast-Food Breakthrough

Remember Ray Kroc, the happy Model T driver from chapter 5? Fast-forward to 1954. The spread of the automobile and the growth of reliable roads connecting cities transformed where people lived and how they lived. It gave rise to what we now call the suburbs and created new phenomena, like the drive-in restaurant. With its temperate climate and extensive highway system, California became the natural center of the drive-in.

Customers, typically teenagers, would drive to a restaurant, order food, and wait for it to be delivered to their cars. Restaurant owners competed by offering ever-more-elaborate menus and ever-more-creative ways to deliver food to cars, such as servers (then overwhelmingly female) on roller skates.

In 1954, Kroc was a food service equipment salesperson with national rights to sell five-spindle Multimixer machines that made milkshakes, five at a time. Kroc, whose colorful autobiography, *Grinding It Out*, bounces between business sound bites ("nothing recedes like success," "if you think small, you'll stay small," "persistence and determination alone are omnipotent") and his three marriages, had heard of a San Bernardino, California, hamburger stand that was doing something remarkable.[1] He heard stories of fast, reliable service of delicious food. The menu had only nine items, but customers loved it. Kroc went to visit, with visions of learning how to boost Multimixer sales dancing in his head.

Kroc saw sharply dressed workers and an immensely crowded parking lot. He talked to repeat customers and heard a kind of deep loyalty that struck him. "I could feel myself getting wound up like a pitcher with a no-hitter going," he recalled. "This had to be the most amazing merchandising operation I had ever seen!"[2]

At dinner with the brothers who ran the restaurant, he struck a deal. Kroc would become the master franchisor of the brothers' operations. He would start by opening his own store in Des Plaines, Illinois, and then had the right to license the store to other operators. He could collect a $950 fee from each franchisee ($11,000 today) and a service fee of 1.9 percent of each restaurant's sales. Kroc's company—officially formed on March 2, 1955—could keep 74 percent of that total (1.4 percent of sales), with the remainder going to the brothers.

In 1960, Kroc named his company McDonald's Corporation. Over the next decade, Kroc and his team laid the foundation of a global powerhouse, all thanks to a special sauce—and not the kind that goes on Big Macs.*

---

* In 1967, the Big Mac debuted in a McDonald's in Pittsburgh. The specific recipe for the Big Mac's special sauce is secret. Key ingredients include sweet relish, various forms of sugar, egg yolks, paprika, turmeric, and—everyone's favorite—soy lecithin. A five-star-rated recipe on the Daring Gourmet website features mayonnaise, dill pickle relish, ketchup, mustard, and paprika. See www.daringgourmet.com/big-mac-sauce. And now you know!

A Brief History of McDonald's

In 1937, brothers Richard (Dick) and Maurice McDonald opened a stand in San Bernardino selling hot dogs and shakes. Three years later, they opened a six-hundred-square-foot restaurant on the corner of Fourteenth and E Streets in San Bernardino. The restaurant had a nontraditional octagonal shape, which exposed the entire kitchen to the public. It sold twenty-five items, employed twenty carhops, and had capacity for 125 cars. The restaurant was a success, with $200,000 in sales ($4.4 million today). The brothers split $50,000 ($1.1 million) in profits.

But the business grew increasingly challenging to run. Strong growth after World War II led to increased employee turnover. As John Love notes in *McDonald's: Beyond the Arches*, "Thanks in part to their teenage clientele, the turnover of eating utensils was as bad as the turnover of carhops. Paying the tab to replace stolen or broken flatware ran counter to their New England thrift ethic."[3]

The brothers innovated. In December 1948, they introduced their Speedee Service System. The idea was radical simplification. Shrink the menu to four prepared items (hamburger, cheeseburger, fries, and milkshakes) and five other drinks (Coke, root beer, hot coffee, orange drink, and milk). Standardize the procedures for cooking and delivery. Remove the frills of carhop delivery, providing counter service instead. The experiment was a hit, and by 1950, the McDonald brothers flashed a sign that would become iconic: "Over 1 million sold."

In 1952, McDonald's made the cover of *American Restaurant Magazine*. The McDonald brothers decided to start licensing the concept to other would-be restaurant owners. The first licensee was Neil Fox in Phoenix. The brothers decided Fox's site should be a prototype for future licensees. They hired a local architect named Stanley Meston, who designed a red-and-white-tiled rectangular building with a sharply slanting roof that became a noteworthy relic of the era. The brothers

Figure 8-1

An early McDonald's restaurant in Des Plaines, Illinois, around 1955

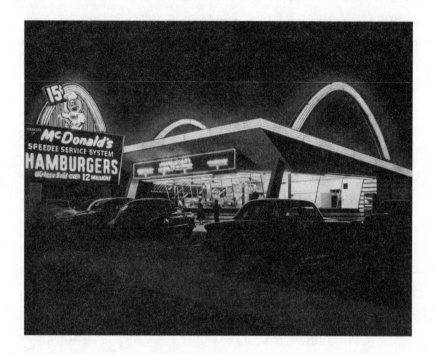

Source: *Neon Arches*, by Hulton Archive/Getty Images, 1955.

wanted an instantly recognizable, memorable beacon for the restaurant. Dick McDonald asked George Dexter, who ran a neon sign company in California, for design help. Ask a neon sign company owner to create a beacon, you are going to get . . . a giant neon beacon. Gaudy? Gauche? Perhaps. But there was no doubt that the two sweeping neon-yellow, gleaming curves forming what resembled an *M* drew attention and would indeed become a visual cue recognizable around the world. The Golden Arches were born (figure 8-1).

Now, one final question. How, precisely, should a new kitchen be designed from scratch to optimize the Speedee Service System? The McDonald brothers provided a world-class demonstration of what in today's world would be called rapid prototyping (and would serve as a wonderful visual moment in the movie *The Founder*, starring Michael Keaton). Love describes the remarkable process:

> *They had a brainstorm. They drew the outline of the new kitchen on their home tennis court, and after closing one night at Fourteenth and E, they invited the night crew over to go through all of the hamburger assembly motions. As the crew members moved around the court making imaginary hamburgers, shakes, and fries, the brothers followed them, marking in red chalk exactly where all the kitchen equipment should be placed. By 3:00 a.m., the tennis court markup was completed, and for a fraction of the cost of conventional design work, the brothers had a detailed kitchen layout.*[4]

We've seen this before. Find a low-risk way to approximate an idea. Try it. Learn. Adjust. Robert Boyle did it in his forty-three experiments on air pressure. Julia Child did it when trying to create the optimal recipe for baguettes. And the McDonald brothers applied it to a key operational uncertainty.

McDonald's wasn't the first franchised food service chain; that was the A&W Restaurant chain created by A&W Root Beer founders Roy Allen and Frank White in 1924. Nor was McDonald's the first carhop drive-in; that sort of restaurant predated McDonald's by six years. And McDonald's wasn't the first restaurant chain to get to scale. Dairy Queen had twenty-five hundred outlets by 1948, the year the brothers introduced the Speedee Service System. But under Kroc's guidance, McDonald's was the first to achieve scale and success at levels previously unimaginable.

And the special sauce wasn't just the food. It was the unique way McDonald's (1) created value for its customers, (2) delivered that value to those customers, and (3) captured a portion of that value for itself. Today we'd call those actions the three components of a business model.

Create Value for Customers

From the 1948 creation of the Speedee Service System, McDonald's was obsessed with standardization and simplification. Consider the french fry.* A simple dish, right? Simple to make an average french fry with the right equipment at home. Complicated to make a delicious french fry that serves hundreds of millions of people a day. One of Kroc's memories of his first visit to McDonald's was the care that the brothers paid to the simple spud. It's a long quote, but it showcases the care and attention paid, even in 1954:

> *The McDonald brothers kept their potatoes—top quality Idaho spuds, about eight ounces apiece—piled in bins in their back warehouse building. Since rats and mice and other varmints like to eat potatoes, the walls of the bins were two layers of small-mesh chicken wire. This kept the critters out and allowed fresh air to circulate among the potatoes. I watched the spuds being bagged up and followed their trip by four-wheeled cart to the octagonal drive-in building. There they were carefully peeled, leaving a tiny proportion of skin on, and then they were cut into long sections and dumped into large sinks of cold water. The french-fry man, with his sleeves rolled up to the shoulders, would plunge his arm into the floating schools of potatoes and gently stir them. I could see the water turning white with starch. This was drained off and the residual starch was rinsed from the glistening morsels with a flexible spray hose. Then the potatoes went into wire baskets, stacked in a production-line fashion next to the deep-fry vats. A common problem with french fries is that they're fried in oil that*

---

* A writer's debate. Capitalizing the *F* in *French fries* seems weird. But I did in the first few drafts. Eagle-eyed copyeditor Patricia Boyd said no. I fought a few changes . . . a writer always does . . . but was swayed by Patricia's invocation of *Merriam-Webster's Collegiate Dictionary* (11th edition).

> has been used for chicken or for some other cooking. Any restaurant will deny it, but almost all of them do it. . . . There was no adulteration of the oil for cooking french fries by the McDonald brothers. Of course, they weren't tempted. They had nothing else to cook in it. Their potatoes sold at ten cents for a three-ounce bag, and let me tell you, that was a rare bargain. The customers knew it too. They bought prodigious quantities of those potatoes. A big aluminum salt shaker was attached to a long chain by the french-fry window, and it was kept going like a Salvation Army girl's tambourine.[5]

Kroc and his team did indeed lavish attention on french fries. In February 1956, Kroc met Fred Turner, who with three family members initially paid $950 to become an early franchise owner. While the family partnership was looking for a suitable site, Turner worked in Kroc's store in Des Plaines and sold Fuller Brush supplies door-to-door to support his wife and young daughter. When the family partnership fell apart, Kroc hired Turner as an assistant manager in a newly opened store in September 1956. At the end of the year, Kroc asked him to join the central McDonald's team to help new franchises open and operate their store.

Turner threw himself into the job, obsessively experimenting with different approaches to improve the french fry. He would bring thermometers to calibrate the temperature of driers and then armed his field representatives with hydrometers. In 1957, the husband of June Martino, McDonald's corporate secretary, got permission to open a formal lab. The lab team developed a "potato computer" to optimize the frying cycle for potatoes. The company estimated it spend $3 million in its first decade under Kroc to improve the french fry. That amount—about $35 million in today's terms—was a lot of money for a young company still sometimes struggling to meet its monthly payroll.

It was all part of a commitment to standardizing the fast production of quality food. Turner took the fifteen dos and don'ts that formed the basis of the Speedee Service System and turned it into a 38-page manual. By 1958, it was a 75-page manual. By 1968, it was a 750-page homage to disciplined operations. The original manual had a foreword by Turner:

> *Herein outlined is the successful method. High-volume McDonald's will always have an old mother hen cluck-clucking around from one corner to another, never being satisfied.* YOU MUST BE A PERFECTIONIST. *There are hundreds and hundreds of details to be watched. There isn't any compromising. Either, (A) the details are watched and your volume grows, or (B) you are not particular, not fussy, and do not have a pride or liking for the business, in which case you will be an also-ran. If you fall into the "B" category, this is not the business for you.*[6]

In early 1957, Turner started visiting stores, grading each on quality, service, and cleanliness. *Quality, service, and cleanliness* remains an industry-standard term today.

Deliver Value to Customers

McDonald's could only succeed and scale if its franchise owners succeeded and scaled. One of the fundamental tensions in a franchise model is between a short- and long-term orientation. Prior to McDonald's, franchise owners tried to get as much as possible, as quickly as possible, from franchisees. That meant charging high up-front fees. To justify those fees, a franchise owner could give an operator significant operating freedom or exclusivity over a large geographic territory. If they didn't run the franchise well, well, the franchisor already had their money. Many franchise owners also forced operators to purchase products or equipment.

McDonald's followed a fundamentally different approach. The company was picky about its franchise owners and only allowed people to start with a single franchise. It offered intentionally narrow geographic locations, giving exclusivity around a one- or two-mile radius. It didn't force operators to purchase products or equipment. The deal negotiated by the McDonald brothers meant low franchise fees and profits largely staying with the local operator. This long-term, collaborative orientation—balanced by McDonald's quality standards—was a key driver of McDonald's expansion.

"The independent-mindedness of our operators prevents regimentation," Turner noted. "While they stick to the basics of the system, they zig and zag by making refinements and changes, and everyone benefits from their willingness to zig and zag. The system deals with setting uniform standards, but regimentation? No way!"[7]

How the McDonald's system developed new products shows the power of the symbiotic connection between franchise owners and the central corporation. Despite Kroc's entrepreneurial moxie, he had a stunning inability to successfully develop new menu concepts. Given his connection to the milkshake business, he had lots of ideas for desserts: brownies, strawberry shortcake, and pound cake. None of them stuck. In many parts of the United States, people didn't eat meat on Friday, for religious reasons. With this in mind, Kroc developed a sandwich called the Hula Burger: a grilled pineapple and two slices of cheese. As delicious as that sounds, it flopped. Kroc also had an idea for a roast beef sandwich. Another dud.

No matter. The fortunate find ways to flip failure into success, and Kroc did just that. McDonald's had four big new concepts in the 1960s and 1970s: Filet-O-Fish, Big Mac, hot apple pie, and Egg McMuffin. Each idea originated with a franchise owner.

For example, in the early 1960s, franchise owner Lou Groen in Cincinnati was in a fierce battle with a local chain restaurant called Frisch's. He particularly struggled on Fridays because he had no viable meat alternative for religiously minded customers. Customers flocked to Frisch's for its halibut sandwich.

"A lot of customers felt that if I didn't want their business on Friday, then I could forget about getting it on other days of the week," Groen said. "I had to have a damn fish sandwich."⁸ He developed an idea to dip halibut in pancake batter, deep-fry it, and serve it on a bun. McDonald's corporate executives approved a local market test. The sandwich did well, but its preparation was painstakingly manual. How to simplify the process?

McDonald's put out a call to potential suppliers. Bud Sweeney from Gorton Corporation agreed to collaborate to develop a scalable fish sandwich (pun intended). It took a year of careful testing, a switch from halibut to cod, and the introduction of melted cheddar cheese before the Filet-O-Fish was launched nationally.* Love's summary of the transition from local to national in *Behind the Arches* shows how McDonald's didn't just create a restaurant concept; it created a system:

> *The introduction of fish brought McDonald's unique marketing partnership full circle. The idea had originated with a single franchisee reacting to his local market. It had been converted into a national product by McDonald's Corporation, which changed it to fit the requirements of the operating system it had designed. And it was manufactured by an independent supplier showing the same commitment to McDonald's needs that its employees and franchisees had. It amounted to a near-perfect utilization of a franchising system.*⁹

Kroc agreed. "The company has benefited from the ingenuity of its small businessmen while they were being helped by the system's image and our cooperative advertising muscle."¹⁰ You go, Ray!

---

* In early 2024, McDonald's CEO Chris Kempczinski visited the Tuck School of Business at Dartmouth College. He said one of his favorite dishes is the Filet-O-Fish, double ketchup, no tartar sauce. Another favorite is the Quarter Pounder with Cheese. He said he eats McDonald's food twice a day. Why only twice? "Variety is the spice of life," he said.

A final component of how McDonald's ensured the consistent delivery of value was the company's investment in training. In 1961, it launched Hamburger University. A grandiose name with a modest start: the first Hamburger U was a room in the basement of a McDonald's restaurant. Over time, however, it lived up to its name. In 1968, McDonald's created a $500,000 ($4.4 million in modern terms) facility with two classrooms. In the early 1980s, it built a $40 million ($125 million) facility that had capacity for 750 people and offered thirty-six hours of accredited courses.

Capture Value for Yourself

How did McDonald's make money in its early days? The deal Kroc struck with the McDonald brothers made it hard for the central corporation to grow through fees and restaurant sales, especially as it was investing in underlying systems. Through 1960, its restaurants had sold $75 million worth of food, but McDonald's earned a measly $159,000 in aggregate profits over the six years of its existence. Its net worth was just $95,000. Recall, the McDonald brothers alone earned $50,000 in a single year in the 1940s!

Kroc was also spending significant time in discussions with the brothers. Their formal agreement was quite loose, leaving room for interpretation, discussion, . . . and frustration. Kroc had successfully expanded the original deal from ten to ninety-nine years, but that wasn't enough. He sought a way to buy the brothers out.

"This was partly for personal reasons," he wrote in his autobiography. "Mac [Maurice] and Dick were beginning to get on my nerves with their business game playing. . . . But the main reason I wanted to be done with the McDonalds was that their refusal to alter any terms of the agreement was a drag on our development."[11]

Kroc eventually negotiated a $2.7 million buyback in 1961 ($28 million today), giving McDonald's substantially more freedom to share in the value created by its franchises.* However, it had to take on debt to finance the transaction, which further limited its operating freedom. And it still believed fiercely in creating conditions that allowed the franchise owners to thrive. How to square the circle?

Enter Harry J. Sonneborn. Kroc had met him in his Multimixer days, when Sonneborn was the vice president of Tastee-Freez, a soft-serve ice cream chain.** Sonneborn joined McDonald's in 1955. A University of Wisconsin dropout, he was a foil to Kroc. Kroc was outgoing, liked operations, and loved hamburgers. Sonneborn was introverted, liked finances, and didn't particularly care about hamburgers.

He did care—a lot—about making money. And he developed the model that allowed McDonald's to break out of its stasis: make money through real estate.

McDonald's created a subsidiary called Franchise Realty Corporation. It would find locations for new stores and enter into a lease agreement for the space. It would then create with franchise owners a twenty-year sublease that had a 40 percent margin for McDonald's.

"I didn't justify it to anybody," Sonneborn said. "That was the deal—take it or leave it. I never went through the exercise of telling the franchisees what their rental factors were. This was the rent."[12]

Consistent with McDonald's theme of creating conditions for franchises to thrive, the company structured its real estate deals so that owners paid a minimum base rent and then a percentage of sales once the franchise crossed a sales threshold. In other words, McDonald's only did well when the franchise did well, and vice versa.

---

* The McDonald brothers demanded that they retain ownership of their original store. Kroc agreed to the demand and then opened up a rival store across the street. Genius? Spiteful? Good business? You make the call.

** I love little historical quirks. There was a phase where company logos were their names in a font known as Spencerian script—look at Coca-Cola, Ford, and Johnson & Johnson. The 1940s and 1950s loved its double *e*'s: Speedee Service, not Speedy Service. Tastee-Freez, not Tasty-Freeze.

"After a time," Kroc said, "we began realizing substantial revenues from the formula, and we could see that we were merely nibbling around the edges of this huge hamburger frontier we were exploring."[13]

The Power of Business Models

So, the world's most famous hamburger restaurant became the world's most famous hamburger restaurant because it figured out a successful formula for extracting rental income from its franchise owners without constraining their ability to grow. The McDonald's business model—standardize through the Speedee Service System, share value with franchise owners and suppliers, and lock in profit with the real estate model—was the engine of its disruptive growth.

The story makes it sound simple. But it took fifteen years of trial and experimentation before those components clicked. And despite all this taking place in plain sight, no one could match McDonald's. Part of the power of a business model is that it is not a *thing*. It is a *system*. Create value. Deliver value. Capture value. Success requires all three pieces, working together in harmony.

In the early 1960s, a Burger Chef franchise owner looked at how that system had come together with admiration.

> *When a guy came out of Hamburger University, he was convinced he was the best damn restaurant operator in the world, and he could conquer anybody. Burger Chef had similar rules on quality, service, and cleanliness, but we just didn't get it done as effectively as McDonald's. There are no secrets in this business—a hamburger is a hamburger. I don't think we had the dedication to quality that they had.*[14]

We can look back and see this system-level approach as a key part of the pattern of disruptive innovation. For Henry Ford to deliver his car for the multitudes, he had to have the car itself, the production system, wages to reduce turnover, a public relations blitz, and much, much more. Gutenberg's printing press wasn't a single innovation; it was an intertwined system of innovation. When you get the system right, you create a powerful growth engine.

In 1959, Sonneborn became McDonald's CEO, with Kroc remaining actively involved as an owner and a board member. In 1965, McDonald's went public. Two years later, the battle between Kroc and Sonneborn (hamburgers versus real estate) boiled over, with Sonneborn resigning and Kroc returning as CEO. Turner took over as CEO in 1973. The next decades featured relentless global expansion. Canada. Germany. Japan. China. Millions served became billions served.* The power of a disruptive business model.

McDonald's Echo: Netflix's Rise

Business model wasn't a popular term when Kroc and his team worked their magic. Today, there are various tools and frameworks to describe a business model.** Just as it took decades for McDonald's to truly perfect its business model, it took decades for the very concept of a business model to become popularized. The term began appearing in the popular press as the commercial internet exploded in the 1990s. The success of Netflix shows how business models continue to be the special sauce of disruption.

---

* When visiting family in Washington, DC, in July 2024, we passed a McDonald's whose sign read "80 billion served."

** My two favorite guides to business models came out in 2010: *Reinvent Your Business Model*, by Innosight cofounder Mark W. Johnson (originally called *Seizing the White Space*) and *Business Model Generation*, by Alexander Osterwalder and Yves Pigneur. My *create, deliver,* and *capture* shorthand came from a consulting project with Procter & Gamble, who features in the next chapter.

Netflix's stated origin story is that Reed Hastings, its cofounder, forgot to return *Apollo 13* to his local video rental store and had to pay exorbitant late fees. In frustration, he founded a company that got rid of late fees.

Like most origin stories, it is nice but not completely accurate. You see, when Netflix was launched in 1998, key pieces of its business model mirrored those other video stores. Yes, you went online and clicked to rent a movie; that was a different way to create value. Yes, the movie came to you via the mail; that was a different way to deliver value. But how did Netflix capture value? In the early days, if you returned a movie you rented from Netflix late, you paid late fees. There was no innovation there. Netflix followed the model the market leader followed. Let's remember that company.

When I was a teenager, you would go to your neighborhood Blockbuster if you wanted to rent a movie. Say it with me, loudly. *Blockbuster.*

The company's stores were small, which meant they could be in dense areas that lacked space for bigger stores. The positive of that design choice was that going to the local Blockbuster was a short, convenient trip for customers. However, a small store created a conundrum. Blockbuster made a promise to its customers. When you walked in the door, you would see a certain type of movie. What type of movie was it?

Of course, it was a blockbuster. It's the name of the store, after all!

Small stores couldn't have infinite inventory. That meant if customers held on to a movie for too long, new customers would find bare shelves. How could you make sure that people would return a movie they had rented? Easy enough. You charge late fees. Not only are the fees a great way to allow you to deliver against your brand promise, they allow you to make a lot of money.

So, in 1999, Netflix added the final piece of its disruptive puzzle: it changed the way it captured value. It shifted from single movies with late fees to a monthly subscription model. Streaming, of course, would then follow. What was Blockbuster to do? Netflix had gone right after the core of the then leader's business model. Blockbuster could have bought Netflix for $50 million in 1999. It didn't. It struggled. It failed.

Netflix would go on to keep innovating its business model. It created value for customers by allowing them to binge on new releases. It delivered value by developing sophisticated algorithms that allowed it to predict what shows and movies would hit. Netflix then used that data to develop compelling original content. Through increasingly sophisticated pricing models, the company captured value. Netflix has had its stumbles, for sure, and by the mid-2020s, it has become to some degree a victim of its own success, as content owners like Disney and Warner Media have decided to create their own streaming services to compete with Netflix. But it had a beautiful twenty-year run powered by that special sauce of business model innovation.

And if you are curious, *The Founder* is indeed on Netflix, at least as of March 31, 2025. Google (whose disruptive business model I touched on in the last chapter) tells me that it is also available to rent on Amazon Prime (one of two major disruptive business models launched by Amazon in the aughts; the other features in the next chapter). The movie has 674 ratings, with an average score of 4.7 (out of 5.0). *Julie & Julia*, a 2009 movie starring Amy Adams and Meryl Streep (playing Julia Child), is on various platforms and also has a rating of 4.7.

Business model innovation at work. That's enough of that. I'm off to get a McFlurry.*

---

* I love McFlurry soft-serve desserts. My New Year's resolution one year when I lived in Singapore was to try at least twelve different McFlurry flavors in a year. I think I hit fifteen that year. The benefits of global travel!

9

Oh Crap!

Pampers Disposes of Its Competitors

In 1954, business was good for Procter & Gamble. Some 95 percent of homes had at least one P&G product. The company had an admirable track record of driving disruption. In the 1880s, it was Ivory, the soap that floats. In the 1910s, it was Crisco, the first shortening made with vegetable oil. In the 1940s, it was Tide, the "washday miracle."

At the time, P&G was putting the final touches on Crest, a fluoride-based toothpaste that dramatically improved oral hygiene.* And still, the company was hungry for more. As the postwar baby boom and economic expansion provided further economic tailwinds, P&G established an Exploratory Development Division, headed by Victor Mills, a researcher who had previously played a pivotal role in both Crisco and Tide.

---

* The development of fluoride-based toothpaste is a great story, featuring an accidental discovery by a dentist in Texas about what was causing the teeth of local residents to turn black. It hit the cutting-room floor. As always, if you are interested, you know where to find me.

Mills eyed paper-based products. P&G knew a bit about the market because one of its subsidiaries processed cotton linseed into cellulose pulp. Experts, however, doubted that P&G could ever be a big player in the paper-products market. Scott Paper estimated that it would cost P&G $1 billion ($11 billion today) to mimic Scott's capabilities in the market, so Scott assumed its competitor wouldn't do so.*

Wrong.

In 1956, Mills commissioned a team to begin working on a paper-based disposable diaper. When the fruits of the effort—Pampers—became the first brand in P&G history to cross $10 billion in annual revenues in 2012, the UK branch of P&G paid homage to Mills's central role in its development:

> P&G Researcher Vic Mills became frustrated with changing his newborn grandson's cloth nappy. The process was messy: The nappies fit poorly, soaked too quickly and needed to be laundered too frequently.
>
> Mills decided there must be a better alternative to the cloth nappies, and, in 1956, asked his Research and Development colleagues to work on developing the first high-quality, affordable disposable diaper.[1]

The Pampers story is eerily familiar.

In my career, I've seen hundreds of ideas with disruptive potential. The industries they target, the technologies they use, the teams behind them, and the business models they follow are all distinct, but they share one important commonality. Every idea, and I mean every idea, starts out partly right and partly wrong. What separates success from failure isn't the quality of the idea; it is how the team handles the inevitable twists and turns in the path to success.

---

* Yup. It's a bit weird for me to read a sentence that talks about "Scott's capabilities."

You see, Pampers was far from an overnight success. Mills and the team had to fail, fail, and fail again before they found a path to success. And two modern efforts to drive disruptive growth under the Pampers brand followed a similar pattern. Disruptive success, it turns out, requires patient perseverance and a dose of luck.

A Brief History of Pampers

In the 1830s, William Procter, an English storekeeper turned candle-maker, and James A. Gamble, an Irish soap maker, emigrated to Cincinnati. They married sisters and, in 1837, formed a company. With its access to the Mississippi River, Cincinnati was expanding as an emerging industrial city. Its growing stockyards led to an ample supply of tallow, a key ingredient for soaps and candles.

P&G sold candles to the Union Army, but that business began to stagnate as consumers switched to oil-based lanterns. The transition from candles to soap serves as the origin story of what would become a powerful success formula.

In 1863, James A. Gamble's son James Noble Gamble described how a customer "spoke well of the floating soap I had given him."[2] The product, at the time unbranded, was made not of tallow, grease, or lard, but of vegetable oil. Procter & Gamble had formulated a product that had comparable quality to premium-priced soaps but was relatively affordable. The company diligently worked to perfect the soap. Harley Procter, a second-generation company leader, credited divine inspiration for the product's brand name. He was in church reading aloud. The congregation read Psalms 45:8: "All thy garments smell of myrrh, and aloes, and cassia, out of the Ivory places whereby they have made thee glad."

Ivory soap was launched in 1878 as an upscale mass-market product. Harley Procter led its marketing, taking out an ad describing it as

"99 44-100 per cent pure," and the "soap that floats." P&G built a huge manufacturing facility near the emerging railroads, and it was off to the races.

This success formula—technical expertise, brilliant branding, manufacturing at scale, and a bolt of luck—was repeated with synthetic laundry detergent. Consumers used the company's soaps to clean clothing, but the process was laborious and largely ineffective. In the book *Rising Tide,* Davis Dyer and his coauthors describe how the process that led to Tide was not linear: "The company assembled the key technologies in fits and starts, by way of fortuitous accidents, vexing dead ends, and slow laborious work. Actually, P&G almost missed the opportunity. The breakthrough grew out of a seemingly dead project—a line of research nearly everybody had given up on."[3]

In 1931, P&G's Robert Duncan was visiting the IG Farben research laboratories in Ludwigshafen, Germany. On the drive back to the hotel, a company representative casually told him about a development that was "conceptually interesting" but surely of no commercial interest to P&G. Facing a soap shortage during World War I, chemists in IG Farben had used bile from slaughtered cattle as a wetting agent. Chemists isolated the ingredient, synthesized it, and sold it under the name Igepon.

It took a decade of development work—some of it through an underground project called Project X—before P&G turned that ingredient into a solution with significant market potential. Technical leadership, including Mills, then an up-and-coming R&D leader, supported developing a pilot plant. Initial success in 1945 led to the company's aggressive push to bring the product to market. It launched the product in 1946. P&G advertised Tide as the "washday miracle" that promised "oceans of suds" that were "cleaner than soap." Tide's iconic bull's-eye logo became a regular feature on the "soap operas" P&G had pioneered as a novel form of advertising. Tide developed a huge lead over competitive products from companies such as Colgate and Lever, a lead it maintains eight decades later.

Over the next decade, Mills sharped his reputation within P&G as a top-flight technical scientist by shepherding the development of continuous soap hydrolyzers, spray-dried synthetic detergent, and fatty alcohol synthesis.* He also led innovative projects, such as an improved cake mix that pushed Duncan Hines to a lead position among baking mixes (P&G sold the Duncan Hines brand in 1997). He was ready for his seminal contribution.

Targeting Disposable Diapers

Two years after taking over the Exploratory Development Division, Mills chartered a new R&D team, led by Bob Duncan, the grandson of Tide pioneer Robert Duncan.

With three million babies being born annually in the United States, disposable diapers targeted a large and growing market. As washing machines, refrigerators, and microwave ovens became more ubiquitous, demand for convenience-related products grew.

By the mid-1950s, 80 percent of households with infants had disposable diapers, which cost about nine cents each. Made by Johnson & Johnson, Chux, and Parke-Davis, these diapers had severe limitations. Poor fit led to leaks; poor material made the diapers crumble, leaving paper on the baby's skin. Disposable diapers were hard to . . . dispose of. Parents would put plastic pants around the disposable diapers, causing rashes and other discomforts for the baby. So, although disposable diapers were in everyone's home, they were only used in case of an emergency when the family was traveling. They constituted less than 1 percent of all diaper changes. Commercial diaper-cleaning services made up about 5 percent of all diaper usage, with cloth diapers that consumers laundered themselves making up most of the market.

---

* I don't know what these are, either, but they sound cool. These are listed in a *Research-Technology Management* article by Harry Tecklenburg in 1990.

To break open the category, P&G needed to come up with a product that performed better than what was on the market but that remained affordable. The specific design goal was for a diaper that held as much moisture as two cloth diapers and would cost around fifteen cents per diaper. Project leader Duncan was the father of twins who were about to turn two years old. He turned his house into an active laboratory for his work. He later described the experience:

> *Now there are some folks that diaper to keep the baby dry and clean. There are others that think there are better things in life to do than change diapers. Betty and I were from the latter school. We would encase the infants in plastic pants and put as many diapers—two or three—as we could get around them, as infrequently as possible consistent with catastrophe. . . . I brought home some laboratory scales, which I put on a bathroom shelf, and I started weighing diapers, particularly overnight. We're talking metric now—250 grams overnight, or 300, was not unusual. It was in this range. The double cloth diapers did not do well. Now, we're talking a good amount [of] moisture. I was trying to get some quantitative data on absorptive load requirements and effectiveness of containment.*[4]

Duncan's home experiments, coupled with fieldwork, showed that diapers had to deal with high variation, in terms of both infant waist size and the amount of urine produced (the amount can vary by 50 percent from one day to the next).

That was the first of many surprises on P&G's path to success. Making the complicated simple and the expensive affordable is hard work, with plenty of failures along the way.

P&G first tested its disposable diaper in the market in Texas in 1958. The diaper had an absorbent insert that was folded to form a comfortably fitting bucket that absorbed moisture. Sales were so low that some P&G leaders suggested abandoning the effort. Maybe the Texas heat biased parents against the separate plastic pants that encased the diaper, advocates argued. In a wonderful retrospective in 1990, P&G veteran Harry Tecklenburg said that perhaps it was not surprising that mothers didn't want to "subject their infants to the rash-producing Turkish bath that comes with plastic pants."[5]

P&G didn't give up.

A few months later, it developed a single-piece diaper with no pants. It created a sheet that sat between the baby and the absorbent material, allowing fluid to go through the absorbent material but not come back. The complexity of folding different parts of the diaper meant it took a full year to create a functioning production line, so the team assembled thirty-seven thousand diapers by hand.

Again, the product struggled in test markets. Only a third of mothers said the diapers were superior to cloth ones, and consumers were willing to continue to use the new diapers only if they were free.

P&G didn't give up.

The third trial run was in Illinois. P&G had a ten-cent diaper—well below its original target price but still more expensive than competitive products—that had a plastic back sheet that went up the sides of the diaper to improve containment and a hydrophobic top sheet designed by Duncan to trap moisture in the diaper's middle layer.

The product again struggled to gain traction, with people put off by the ten-cent price. Sales were about a third of P&G's targets.

A Fishing Trip Leads to the Diaper Wars

Perhaps, the team thought, we were wrong to assume that success would come with a product priced at a modest premium over existing disposable diapers. The development team members decided to set a new target. To do that, they decided to . . . take a fishing trip.

During that trip, they decided that 6 to 6.5 cents was the magic number. "With a precision completely unjustified by the scantiness and spread of the data," Tecklenburg recalled, "we established the organizational goal of 6.2 cents per diaper."[6]

Hitting that target pushed P&G's engineering capabilities to the max. A retail price of 6.2 cents required a manufacturing cost of 3 cents. Engineers had to develop new-to-the-world machines and processes that would assemble diapers individually at high speeds and at low costs.

Despite the test market failures, P&G persisted. It launched its product, under the brand Pampers, in Sacramento, California, in 1963, at a price of 6 cents. Success! Almost.

Sales were not quite at company expectations. So, the team added an extra-absorbent product designed to help babies sleep through the night, and it cut the price for its core product to 5.5 cents in its St. Louis launch in February 1964. The team hit its target for launch, but success took longer than expected.

Still, P&G had seen enough to decide to launch nationally (see figure 9-1 for an example of the product's early advertising).

In 1969, John A. Shiffert, the executive vice president of the Diaper Service Industry Association, which represented the four hundred diaper service companies that collectively made $85 million a year, said, "I would be less than honest if I told you that the association is not concerned about the competition presented by disposable diapers."[7] Shiffert was quick to point out the limitations of disposable diapers, citing market research that showed the number one consumer need was to develop an odor-free container for diapers waiting to be serviced.

Shiffert was right to be concerned. P&G didn't get the original design right. It took a fishing trip to figure out the price that would unlock the market. But once P&G got the pieces in place, the company learned that its assumption that the disposable market would be small was wrong. Very wrong.

P&G had assumed that disposable diapers would be a convenience for traveling parents and would capture about 6 percent of the cloth

Figure 9-1

Early Pampers advertisement, 1967

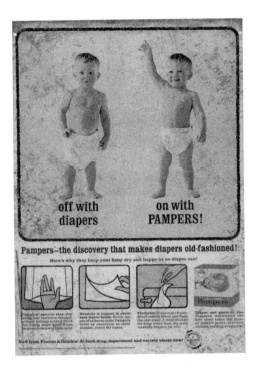

Source: Photo by author.

diaper market. The market significantly exceeded its expectations. By 1970, Pampers had 92 percent of the disposable diaper market. It was the largest brand P&G had ever created in its almost 150-year history.

One key to that success was how Pampers wasn't just a nice-to-have convenience for wealthy parents; it was a must-have enabler for more working-class families. *Rising Tide* details a call P&G received from a woman living in a tenement building: "Without Pampers, the woman said, she was forced to take a pail of soiled diapers down four floors and then walk through an unsafe neighborhood to the coin laundry located two blocks away. The same pattern emerged when Pampers was launched worldwide. Consumers in countries where household washing machines and dryers were less common clamored for the product."⁸

As sales took off, P&G had a new challenge: scale manufacturing, and scale it rapidly. P&G expanded capacity in its plant in Cheboygan, Michigan, and added lines to a P&G paper plant in Mehoopany, Pennsylvania, in 1966. Over the next few years, it opened several new plants, adding a production line each month. In 1969, *Time* said the development effort was "worthy of the creation of a new line of automobiles."⁹ By the early 1970s, P&G could produce four hundred diapers a minute.

It is hard for companies to maintain overwhelming market share—especially in fundamentally attractive markets. Restless innovators seek to find their own disruptive entry paths. Complacency seeps in. Top employees leave. Earning success is hard; maintaining it can sometimes be harder.

In the 1970s, Kimberly-Clark introduced a "technologically superior diaper for the masses" under the Huggies brand. After evaluating different response strategies, P&G chose to leave the lowest end of the market, develop a new product to compete against Huggies, and invest close to $750 million ($2.2 billion today) to develop and launch Ultra Pampers. *Rising Tide* called the move "the largest single construction and start-up project in the company's history."¹⁰ P&G launched the product in 1986 and achieved distribution across the United States in eleven months, a record for the company.

In parallel with the diaper wars, P&G started to pay greater attention to China. The country was in the early stages of what would be a three-decade expansion that would transform it from one of the world's poorest countries to the world's second-largest economy. P&G launched its first probes in Beijing and Shanghai in 1985. In 1991, P&G President John Pepper came to the country for an extended visit. He said success would require three factors: technology that was superior to local offerings, strong relationships with all levels of government with a goal of "making ourselves a company of China," and strong local talent.

At the turn of the twenty-first century, P&G was the clear leader of a big global category. Its diaper business operated at an impressive scale. The company, however, remained hungry for more.

Modern Efforts to Disrupt Again

In 2004, I was part of a team at Innosight advising Procter & Gamble. Our day-to-day client was Dave Goulait, an engineer who had spent more than three decades in P&G's Baby Care Division. Gil Cloyd, chief technology officer, had asked Goulait to "innovate the innovation process." We worked with Goulait and his successor, George Glackin, over the next few years to develop a repeatable ability to incubate and launch disruptive innovations. The results of this work were described in a 2011 *Harvard Business Review* article, "How P&G Tripled Its Innovation Success Rate."

During those engagements, we met two teams pursuing potentially disruptive diaper ideas. Both ideas looked attractive. Both had high disruptive potential. And both had cool code names: Liberty and Joey. But both failed, initially at least. Then both changed course. One of the ideas ultimately succeeded. One did not, at least commercially. Let's look at these two ideas.

The Liberty team, which included Goulait, was working on an extremely low-cost diaper for the Asian market, with a particular focus on China. At the time, existing disposable diapers were too expensive for mass-market customers. During the day, customers would instead put children in split-crotch pants known as *kaidangku*. From a very young age, parents taught babies who needed to relieve themselves to go outside, squat, and let nature do its work. At night, parents would use cloth diapers. The Liberty team's initial hypothesis was that sharply lowering the price would allow parents to convert from *kaidangku* and cloth diapers to disposable diapers.

The Joey team was targeting customers in wealthier markets. From extensive discussion with consumers, project leader Jill Boughton and her team had identified a clear pain point. Imagine you are a parent at home watching your active child happily play. You see the telltale bulge. The child has filled the diaper with urine and needs a change. You sigh. You know that your baby will be upset when you interrupt playtime, and your child will wriggle furiously while you try to change them.

The team's idea was a diaper with an insert called a slip-in. If the diaper was just wet, a parent could easily swap the slip-in for a fresh one without interrupting a current activity or using a changing facility (a messy diaper still required a full change). The team called the product Pampers Change 'N Go.

Some members of the technical community were skeptical that P&G could deliver a viable product. "We got told by senior R&D leaders that there was no way the product could work because it defied the laws of physics," Boughton said. "I guess the laws of physics can be wrong, because the product itself was really, really good."[11]

Like all potential disruptions, Liberty and Joey began their life with imperfections.

The Liberty team (with the ultra-low-cost solution) ran several test markets. It quickly learned that its product wasn't connecting with the market. Price was a barrier to adoption, but it wasn't the only barrier. And people were generally happy with existing solutions. To get to its target low price point, Liberty had to make significant performance trade-offs. Consumers viewed the diaper as a cheap, inferior solution.

Studying its struggles highlighted a different path. *Kaidangku* pants and cloth diapers worked fine when children (and parents) were awake. When children were sleeping, however, and they wet themselves, the urine cooled in the cloth, and they would start to cry. Children couldn't sleep. Parents couldn't sleep.

In *Competing Against Luck*, Clay and his coauthors recounted how Goulait observed a pivotal focus group with people who had tried the Liberty product. "As the moderator worked through the standard protocol of questions—how did the experience go, what was the high point of the week, and so on—one woman's answers caused hearty laughter from the group. What had she said to trigger that response? The translator giggled too. The highlight of this woman's week was renewed intimacy with her husband—three times in that one week."[12] A diaper with greater absorbency allowed the woman's baby to sleep through the night. The woman then said the husband said the diaper was "the best ten cents he ever spent."

What if, the team thought, the consumer benefit pivoted from a cheap disposable diaper to an affordable night's sleep? The team would have to boost the product quality, abandon overly focusing on its price target, and figure out how to communicate a new benefit.

The Joey team (with the slip-in diaper modification) also learned that its product wasn't connecting with the market. The team members brought samples of the product to consumers at home. The team learned that parents believed they had the tools to overcome the challenge of an active diaper change for a struggling child at home. And, in the at-home context, the performance trade-offs that came along with early versions of the Joey solution were unacceptable.

Hope was not lost. The group identified a different situation where parents had a different, potentially more pressing problem. Imagine you are at an amusement park with your child. You see the telltale bulge. You have to stop doing whatever you are doing. And now you have a choice: change the baby in public or use changing facilities that might not have been cleaned in the past month.

The team began running tests in zoos, fitness clubs, and sports stadiums and began selling the product online. Despite virtually no marketing, younger, more technologically savvy consumers found the website and started buying small amounts of the product.

In July 2008, Goldman Sachs got wind of the project. A report noted that the "revolutionary" diapers had received "overwhelmingly positive" consumer results and could "significantly alter category dynamics." The report said that Joey "has the potential to hit the sweet spot as: 1) it provides a unique consumer benefit and is very convenient, 2) a significant amount of the diaper's material costs are removed so it should benefit margin, 3) there is a legitimate 'green' marketing angle and 4) there is likely patent protection."[13] Goldman Sachs estimated that the idea could boost earnings per share by three to five cents, which implied the new product could create billions of dollars of value.

Knowing that one team succeeded and one failed, which do you think is which?

Liberty succeeded. Joey failed.

The Liberty team launched concurrent streams of work to rebuild its diaper around the job-to-be-done of getting a good night's sleep and began a research project with the Beijing Children's Hospital Sleep Research Center. The research found that babies who wore the team's diapers fell asleep 30 percent faster, slept thirty extra minutes a night, and, critically, improved their cognitive development. Given the one-child policy that China then followed, this benefit was a crucial one for achievement-oriented parents. P&G's marketers weaved together the Golden Sleep campaign, and a few years later, Pampers had more than $1 billion in local sales and a leading share of the market. It expanded the low-cost diaper globally, driving further growth.

In hindsight, the Goldman Sachs report highlighting that Joey could create billions of dollars of value was the project's high point. In-market experiments suggested that while consumers liked to *try* Joey, two barriers inhibited the kind of repeat purchase required to build a big business.

First, the circumstance where the need was most intense was just too occasional. Second, parents had developed a reasonable set of work-around behaviors that got the job done. The work-arounds were imperfect, but consumers deemed them better than carrying around an extra set of diapers. Without repeat purchase, the Joey financial model looked increasingly like a fantasy, and the company pulled the plug on the initiative.

In a 2023 interview, Boughton wistfully remembered the project: "In hindsight, I wouldn't have marketed it on the go. I would have started it with special needs children. When we did some work with autistic children, parents were shaking and crying. They said, 'Please make this for our child. I'll pay whatever you want for this.' I've never seen a consumer reaction like that to anything."[14] That path would have presented its own challenges, however, as P&G is built to operate on a massive, global scale, not within narrower niches.

Is the Joey story a failure?

It depends on how you frame it. The team members identified the critical assumptions behind success. They ran focused experiments and staged investment behind those assumptions. The data they generated from those experiments suggested that their key assumptions were wrong. And in light of that data, they decided not to make further investments. They extracted learning from the project, which informed other efforts at P&G.

Through the lens of generating immediate new revenue, yes, that's a failure. Through the lens of learning in a disciplined way, it is a clear success.

It happens.

Pampers' Echo: Amazon Web Services

Modern leaders face intense short-term pressure. One famous study found that 80 percent of chief financial officers would cut investment in R&D or advertising to hit a short-term earnings target, and more than half of these same executives would delay starting a new, value-producing project to meet targets.¹⁵ This, even though a stream of research led by McKinsey & Co. shows that the best way to deliver short-term results is to focus on long-term performance. That approach seems counterintuitive, but the long-term orientation guides smart short-term decision-making and leads to overperformance over time. Too many of today's leaders would not have the patience of the top brass at P&G who kept investing over the fifteen years it took for Pampers to go from an idea to market success.

A new disruptive growth business created by Amazon in the first decade of the 2000s shows the power of a long-term orientation.

In the early 2000s, Amazon had a problem. The young startup had crossed $5 billion in revenues and was scrambling to keep pace with its growth. Its internal technology projects were just taking too long. Matt Round led an effort to streamline and standardize parts of development projects. That led to the creation of a shared IT platform. In 2002, Amazon allowed outside developers to create applications that also ran on the platform. Strong market interest caught the company by surprise and fed into a fateful off-site in mid-2003.

Andy Jassy, who had joined Amazon in 1997 as a marketing manager after graduating from Harvard Business School, had just taken over the fledging web services business. At the off-site, leaders started what Jassy thought would be a short discussion about Amazon's core capabilities. They discussed how Amazon provided a wide selection of goods to consumers and was also good at fulfilling that demand. Sensible enough, given that Amazon was a retail business. Leaders then went deeper, asking *how* Amazon provided selection and fulfillment. The group discussed how Amazon was good at running reliable, affordable, scalable, technology platforms. The inherently low margins of its retail business put pressure on it to be *really* good at those enabling technologies.

"In retrospect, it seems fairly obvious," Jassy said. "But at the time, I don't think we had ever internalized that. And we had a whiteboard in front of us, and . . . we had kind of written on the board that we had a consumer business, and then we had a seller business. And we said, 'Well, is there a developer, B2B [business-to-business] business?'"[16]

As Jassy talked to other leaders in Amazon, an idea coalesced for an "internet operating system." Such a system didn't exist externally, but Amazon had built many of the pieces internally.

"Amazon has always been a technology company at its heart," Jassy said. " . . . We realized we could contribute all those key components of that internet operating system. And with that, we went to pursue this much broader mission . . . to allow any organization or any company or any developer to run their technology applications on top of our technology infrastructure platform."

The idea was to allow external companies to essentially rent, rather than purchase, IT services. Today we call this cloud computing. Not everyone was excited about the opportunity. At the time, cloud computing was a nonexistent market, and like other nonexistent markets, there was no way to put an appropriate size on it. Yes, the IT services industry was huge, but the cloud computing market didn't exist.

Because Amazon's initial solution had limitations in terms of service quality and security, it couldn't serve the largest, most profitable customers.

In their book *Lead and Disrupt*, academics Charles O'Reilly and Michael Tushman describe one skeptic: "When John Doerr of Kleiner Perkins, an Amazon board member, learned of this effort, he wasn't happy, seeing this as a distraction. But [Amazon founder and then CEO Jeff] Bezos ignored him, believing that Amazon had a natural cost advantage in this trillion-dollar market."[17]

Amazon targeted smaller companies that couldn't afford proprietary hardware (echoing how the transistor started in nonconsuming markets). As the business—branded Amazon Web Services (AWS)—expanded, it increasingly served large companies seeking the speed and flexibility of cloud computing.

Amazon launched the AWS business in 2006. A decade later, AWS generated $10 billion in profitable revenue. In 2024, that number had increased by tenfold to more than $100 billion in revenue. Jassy served as the CEO of AWS before taking over as CEO of the overall company in 2021.

In a 2011 meeting with shareholders, Bezos explained the philosophy behind Amazon's approach to innovation. "We are willing to think long-term," he said. "We start with the customer and work backwards. And, very importantly, we are willing to be misunderstood for long periods of time."[18]

The willingness to be misunderstood for long periods is a critical and common component of disruptive success.

The Power of Persistence

As I conducted research for this book, I wondered to which the pace of innovation has accelerated. Work on Ivory started in the early 1860s. Success came about fifteen years later. Tide started in the early 1930s. Commercial success happened in the late 1940s. Work on Pampers started in 1956. It took about fifteen years before the product broke through. The Liberty project took a dozen years of effort before it succeeded. If there's a dramatic acceleration, it certainly isn't evident in those stories. In each case, success required patient persistence over a long period, with setbacks and stumbles on the way.

That raises another question. Did Vic Mills, Bob Duncan, Harry Tecklenburg, and the broader Pampers team succeed because they were skilled? They created a successful product. They pushed prices down. They developed the ability to manufacture at scale. They displayed dogged determination over many speed bumps. These skills and other qualities are necessary ingredients of disruptive success. But are they sufficient? What about good fortune? Pure, good ol' random luck?

There are countless moments in these innovators' stories, as in any story, where things could have gone another way. A top leader could have pulled the plug after the debacle of P&G's disposable diapers in Texas. Something disastrous could have happened with an imperfect product, leading to a lawsuit that shut down the effort. The team could have blown up during the fishing trip or gotten so distracted with the fishing that they never got around to developing a dramatically lower price target. The wrong thing could have happened in the economy, in the world, at the wrong time.

Happy accidents play a role in any disruptive success story. But as Branch Rickey, the general manager who integrated baseball by signing Jackie Robinson for the Brooklyn (now Los Angeles) Dodgers in 1947, liked to say, "Luck is the residue of design."* By running in-market experiments, deeply understanding the consumer, and displaying tenacity, Mills and his team created the possibility for fortune to smile. Research into the science of luck and serendipity (yes, there is such a thing; see *The Luck Factor* by Richard Wiseman and *The Serendipity Mindset* by Christian Busch) backs this up. You can create your own luck by keeping your eyes open and putting yourself in circumstances where good things can happen. And you can savor surprises when they do happen.

There's a relationship between the power of persistence and following seeming surprises. In 1985, Henry Mintzberg and James A. Waters framed the challenge elegantly with their *Strategic Management Journal* article "Of Strategies, Deliberate and Emergent." As the title suggests, the article argues that there are two ways to do strategy. A *deliberate* strategy involves studying, planning, and acting. Do a careful analysis. Talk to experts. Build forecasts. Align key stakeholders. And then execute. This works, Mintzberg and Waters noted in an environment that

---

* Google tells me that poet John Milton originated the quote, and Branch Rickey popularized it. You could also use French scientist Louis Pasteur's quote: "In the fields of observation chance favors only the prepared mind."

is "perfectly predictable, totally benign, or under the full control of the organization."[19]

That's not the world in which most disrupters live. They need to embrace an *emergent* strategy, which involves accepting that you can't know the right strategy from day one. Test, learn, and adjust. The approach is not undisciplined; it is just a *different* discipline. Do your homework but focus less on the answers and more on the assumptions behind success. Find the most critical ones. Experiment. Learn how you are wrong. Adjust. Once the right strategy emerges, pounce on it, shut down the experimentation, and execute.

Does this process sound familiar? It should. It's nothing more than Bacon's and Boyle's scientific method brought to strategic uncertainty.

Every idea in its early days is the same: it is partly right and partly wrong. The challenge is, you don't know which part is which. You can study all you want, but the only way to really figure it out is through disciplined experimentation. That was true for Ivory, Tide, and Pampers and remains true today. Vic Mills and his team patiently followed an emergent approach. They persisted through the negative surprises, using them as lessons to spur improvement, and pounced on the positive ones.

Luck is, indeed, the residue of design.

10

The Ghosts of Bethlehem

Big Steel and the Shaping of an Industry

This story begins sometime in 2001. I don't know the exact date, but at some indeterminate time this year, I first heard Clay Christensen tell a classic story about disruption in steel.

Clay would use a single slide to describe the history of the steel industry. Figure 10-1 is from one of the various presentations that sits on my hard drive. The format mirrors the disruptive line model. It had various market tiers (e.g., rebar, structural steel), showing the quality of steel demanded. Each line drawn for the market tier had two numbers associated with it. The number on the left showed gross margin for a particular tier. The number on the right (under the heading "% of tons") was the percentage of industry volume. So, for example, structural steel had margins of 18 percent and made up 22 percent of industry volume.

Figure 10-1

Clay Christensen's steel chart

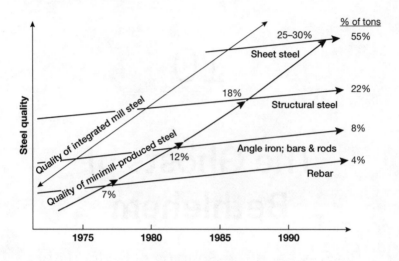

Source: From a May 2015 presentation by Clay Christensen.

I don't know exactly where he got those numbers. A teaching note for a business school case he taught about U.S. Steel has the figure without numbers. Regardless, the story was a staple. I heard it dozens of times. Not only can I tell the story almost word for word, but I can also mirror Clay's pacing, hand gestures, and sound effects.*

Clay would talk about how there are two basic ways to produce steel. The first is to build a big integrated mill. These were complicated and expensive, requiring specialized labor. Bethlehem Steel built the last integrated mill in the United States in Burns Harbor, Indiana. When the company started the project in 1962, it was the largest construction project in the history of the United States. Building the plant took two years and cost $400 million, about $4 billion in today's terms.

---

* I can also do the same for Clay's famous milkshake story, which didn't make it into this book. Catch Clay's version on YouTube. It's good.

The second approach is to build a minimill, which Nuclear Corporation (which would change its name to Nucor Corporation in 1971) decided to do in 1968. Minimills take scrap steel, melt the scrap and iron ore in an electric arc furnace, pour the molten product into a continuous caster, and turn it into finished products. Their small size limited minimills to relatively rudimental end products that didn't require rolling or other forms of finishing. In other words, minimills made basic products at low prices. Classic disruption.

Minimills first began opening in the United States in the 1930s. Operators typically located them in fast-growing Sunbelt cities in the southern part of the country. Those locations had access to growing customers and younger, nonunionized workforces. Nucor's first steel minimill opened in Darlington, South Carolina, in 1968 and cost about $4.5 million (about $40 million today). It had sixty thousand tons of annual capacity, compared with a big integrated plant that would have three to five million tons of capacity.

Nucor's decision to enter steel production didn't catch the attention of market leaders. "It was widely assumed that we could never take the big boys on head-to-head," Nucor's longtime CEO Ken Iverson notes. "Nucor battling Bethlehem Steel would be like a flea going after a rhinoceros. At best we might find room to operate in the fringes of the industry."[1]

Clay deftly described what happened next. Minimills offered a rudimentary solution at rock-bottom prices in the least attractive market tier. So the integrated mills were faced with a choice, he would say. The question was, Should we invest to defend our least demanding, least profitable business, or should we exit that business to focus on better opportunities up-market? The choice seems obvious.

Clay's story had a memorable rhetorical flourish. He would ask what happened when the last integrated mill left a market tier. He would push his large hands down and make a sound best captured as *whoomp*. Prices, he said, collapsed since there was no longer a high-cost producer propping them up. Once the high-cost players left the market, minimills fought against minimills, and prices went down to the cost of the marginal provider.

Clay explained that the minimills' reward for victory was that none of them could make any money. What did they do? They looked up to see a more demanding tier of the market that, if they could enter it, could again earn them juicy margins. Clay said that the integrated mills were faced with a choice: Should we invest to defend our least demanding, least profitable business, or should we exit that business to focus on better opportunities up-market? Again, the choice seems obvious.

The story showed how a series of seemingly rational decisions to invest resources in the most profitable opportunities led integrated mills down a path of destruction.

That's the power of the resource allocation process, identified as a key driver of strategic decisions by Clay's thesis adviser Joe Bower.[*] Resource allocation is the diffuse, difficult-to-tame villain at the heart of the innovator's dilemma. Left to its own devices, it naturally—and sensibly—prioritizes making today better over making tomorrow different. Until there's nowhere left to run.

There's another side to the story. And it involves ghosts. Strap in.

A Brief History of Bethlehem Steel

There is evidence that liquid steel was produced in India in 500 BCE, but it was not "rediscovered" in Europe until 1740. At the time, it was expensive and difficult to produce. The Industrial Revolution and the rise of railroads created a burning need for lower-priced steel. The cast iron used for rails was brittle, inflexible, and affected by extreme

---

[*] I met with Joe in December 2023 to hear his reflections on Clay. Our discussion didn't end up making the book, but it was a wonderfully warm exchange. Which was common whenever anyone reflected on Clay. He was a kind man.

temperatures. Easily cracked rails became an increasingly acute problem for railroad operators. They needed a different solution. Steel was strong, flexible, and less affected by extreme temperatures. However, it was cost prohibitive.

Henry Bessemer discovered a way to remove impurities in pig iron, a key intermediary input into finished steel, in an effective, efficient way. The Bessemer process transformed the production of steel from expensive and burdensome to something that could be done cheaply and effectively. And Bessemer was willing to license the technology to anyone who was interested in it.

One company that took license to the technology was the Bethlehem Iron Company. The company's history traced back to the late 1830s. Its home state of Pennsylvania was rich in the raw materials for iron. Nestled in the Lehigh Valley, the city of Bethlehem is seventy miles away from Philadelphia and eighty miles away from New York. Bethlehem is one of a cluster of cities in Pennsylvania with biblical names. In 1747, it became the first US city to have a decorated Christmas tree. And from 1873 to 1995, it produced steel. A lot of steel.

Bethlehem's Rise

Bethlehem Iron became Bethlehem Steel in 1899. From 1904 to 1945, Bethlehem Steel had only two leaders. It started with Charles Michael Schwab.* Called "the Master Hustler" by Thomas Edison, he lived large but died penniless. While Schwab was working at a grocery store when he was young, the general manager of Andrew Carnegie's Edgar Thomson Steel Works noticed him and hired him. Schwab rose through the ranks and began traveling ten miles daily to deliver an operating report to Carnegie. One day, when Carnegie was late, Schwab sat down at the

---

* There is no relation between steel scion Charles M. Schwab (1862–1939) and financial disrupter Charles R. Schwab (1937–). I suppose if you want to increase the odds of raising a titan of industry, name your child Charles Schwab?

piano and began singing to pass the time. Struck by the vibrant youth, Carnegie made him his protégé and helped him accelerate through the company.

Carnegie's relentless focus was to build mills, to make those mills as big as possible, and to run them constantly. His formula was simple. "Cut the prices; scoop the market; run the mills full...," he said. "Watch the costs and the profits will take care of themselves."[2]

A simple formula. A relentless focus. Others could have seen the same thing, but Carnegie saw it first and acted with, excuse the pun, steely determination.

At a 1900 dinner with a number of bankers, including J. P. Morgan, Schwab gave an impassioned speech about the opportunity to consolidate the steel industry. The next year, Carnegie sold his company to J. P. Morgan (J. P. Morgan the company, not John Pierpoint Morgan the person; they are different but the same), which consolidated the market into a single giant: United States Steel, known as U.S. Steel.

Carnegie himself received $300 million in gilt-edged securities, which is about $10 billion in modern terms. U.S. Steel's capitalization was $1.4 billion ($50 billion today). It was the first time a company's market capitalization crossed $1 billion. At the time, $1.4 billion was equivalent to two-thirds of the money in circulation in the United States. The giant employed more than a quarter million people, more than the US Army and US Navy combined.

Schwab became the president of U.S. Steel. He acquired Bethlehem Steel, sold it to Morgan, resigned from U.S. Steel in 1903, and then retook control of Bethlehem Steel in 1904. Soon after taking full control of Bethlehem Steel, Schwab decided to move the company's headquarters from New York (where Carnegie and U.S. Steel were based) to Bethlehem. "I was convinced that no great steel corporation can be managed from New York," he said. "I [was] determined to reorganize the Bethlehem Steel Co. and have it managed entirely on the grounds."[3] On his seventieth birthday, he called Bethlehem Steel "my own child, the soul of my life. It is my monument that I hope will stand as something worthy to be connected with."[4] Schwab intertwined the company with the city itself, investing in culture, sports, health care, and education, investments that grew as the company grew.

Schwab ran the company until 1916, handing day-to-day responsibilities over to Eugene Gifford Grace, who served as president from 1916 to 1945 and as chair from 1945 to 1957.

A 1924 profile of Grace by the *New York Times Magazine* described him as "a slender, rather tall, nervous type of man, very quick and acute in manner, with a knack for piercing the thoughts of his visitors before they are uttered."[5] Grace oversaw the 1908 construction of the Grey Mill, a mile east of the main blast furnaces in Bethlehem's core plant. The Grey Mill produced the *Bethlehem beam*, a strong, lightweight beam that played a key role in the building boom in cities over the next few decades. He also led the company through a remarkable production stretch in World War II.

Bethlehem grew substantially during World War I. It supplied everything from shrapnel to submarines. The annual earnings of $61 million in 1916 exceeded the total earnings from the eight years before the war. The company produced 60 percent of the guns and 40 percent of the shells used by the United States during the war, and 65 percent of artillery pieces by US Allies.[6]

The Bethlehem beam ushered in the era of skyscrapers. After successful proofs of concept in New York (the headquarters for Gimbel Brothers department store) and Chicago (the Chicago Merchandise Mart and the Chicago Opera House), buildings started going *up*, both literally and figuratively. The sixty-six-story American International Building. The seventy-story Bank of Manhattan Trust Building. Based on Bethlehem Steel. The Golden Gate Bridge. The George Washington Bridge. The Ben Franklin Bridge. The Supreme Court. Based on Bethlehem Steel.

Then came World War II. The day that France and England declared war on Germany in 1939, Grace allegedly told his golfing partners,

"Gentlemen, we are going to make some money."[7] Bethlehem Steel made *a lot* of money. A few weeks after the attack on Pearl Harbor, the US government placed $1.3 billion in orders for bomb casings, armor-piercing shells, gun forgings, airplane parts, and warships. The ships! Bethlehem Steel built 1,121 ships in World War II, including 380 ships in 1943 alone—more than one a day. Employment in Bethlehem itself peaked at 31,523 in 1943, with overall company employment that year of close to 300,000, including 25,000 women.

Bethlehem's Struggles

After World War II, Bethlehem Steel encountered two categories of challenges. The first was labor disputes. Issues with labor had plagued the industry from its early days. A key moment was in 1892 in Homestead, Pennsylvania. Unable to come to an amicable agreement with the Amalgamated Association of Iron and Steel Workers, Carnegie's designate Henry Clay Frick (Carnegie was in Scotland at the time) locked the workers out. Frick brought in nonunion labor to work in the plants, and on July 6, 1892, he also called in a group of three hundred people whom you would charitably call protective agents and, less charitably, call goons. A fight broke out, ten people died, hundreds were injured, the militia was called out, an anarchist shot at Frick . . . and ultimately business restarted with nonunion workers and Carnegie came back to the United States. Who did Carnegie call in to quiet Homestead after the labor dispute? That's right. Charles M. Schwab.

Schwab and Grace fought unionization efforts that encouraged a 1910 strike that required the police to be called in at Bethlehem Steel, a 1937 strike that led to the walkout of eight thousand workers in Bethlehem and the killing of ten (non–Bethlehem Steel) demonstrators by police in the Little Steel Massacre in Chicago. In 1941, another strike led Pennsylvania governor Arthur H. James to declare a state of emergency and bring out troops. Facing the prospect of losing access to wartime contracts, Grace relented, finally recognizing the union. Issues accelerated.

In 1946, eight hundred thousand steelworkers struck and walked off their job.

The year 1952 saw a seven-week strike.

In 1956, a strike ended with Bethlehem Steel agreeing to a controversial clause that said if a task required X workers to perform the task, there would never be fewer than X workers assigned to that task.[8]

In 1959, the company endured a 116-day strike that required intervention by President Dwight Eisenhower. Steel leaders struck a deal that strongly favored workers, setting off a pattern that perpetuated over the next two decades and drove up wages and benefits.

The decrease in flexibility from those agreements coincided with the second challenge: the emergence of the minimills in the late 1960s.

Nucor didn't enter the steel industry to fight against integrated mills.* Rather, it was trying to solve a problem in its profitable Vulcraft business. The division was performing well. It made steel joists, a basic but critical structural construction component. The length of its joists ranged from 10 to 130 feet. Sixty percent of Vulcraft's costs were raw materials, most of which were imported. That left the division subject to the vagaries of the market. What to do? Nucor held a series of open-ended leadership debates that longtime CEO Ken Iverson said were "chaos" as people "waved their arms around and pounded on tables," their faces getting red and veins in their temples bulging out.[9] Their

---

* A blurb on Nucor: The company was originally created to build nuclear instruments and conduct radiation studies. One of its products was the scintillation probe. Yes, that's a real product name. Here's another fun fact: Nucor's history traces back to one of Henry Ford's rivals: Ransom E. Olds. As described in chapter 5, Oldsmobile did pioneering work on the assembly line that Ford perfected. Olds left Oldsmobile in 1904 and launched a company called REO Motor Cars. REO built cars until 1936 and trucks until 1957. Yes, the company's 1915 truck inspired the band REO Speedwagon. In 1955, REO Motor Cars couldn't fight the feeling anymore (couldn't help myself) and merged with Nuclear Consultants, which was ultimately renamed Nucor in 1971.

answer was to backward-integrate. And the group chose the simplest, most affordable way to do so: a steel minimill.

Nucor quickly found a market for its affordable steel, moving beyond selling to Vulcraft to supplying other manufacturers. As mentioned above, it found success in the lower tiers of the market. And that seemed where it was destined to stay. Flat-rolled steel (what figure 10-1 labels "sheet steel") was an attractive market but simply couldn't be created anywhere other than in an integrated mill. Or so everyone thought.

"Minimills have largely passed the phase of displacing tonnage from integrated producer," one CEO said.[10] "We're concentrating on the more sophisticated products where they can't compete, and we've dropped the inefficient lines to them," said another. A leading steel economist agreed, noting, "It will take some time for Nucor to make inroads into flat-rolled steel."[11]

In the movie *Jurassic Park*, the scientist played by Jeff Goldblum quips, "Life finds a way." So do disrupters. What happens is predictably unpredictable. You don't know *how* the disrupter will move upmarket. But you know it is motivated to do so. So, you bet that the disrupter will figure it out in a clever way. Expect the unexpected.* The impossible becomes imaginable becomes plausible becomes prototyped becomes standard practice.

In the early 1980s, Nucor began looking for technology that would enable production of flat-rolled steel in a smaller factory. It specifically looked for thin-slab technology to obviate rolling lines that could stretch for a mile inside an integrated mill. Nucor found a provider in Germany called SMS Schloemann-Siemag, which one Nucor executive called "a little tinker toy operation, a prototype." The American company ended up investing more than $250 million ($650 million today) to use SMS's technology as the backbone of a new plant in Crawfordsville, Indiana.

---

* My thirteen-year-old son, Harry, likes to say this, typically while kneeing you where you don't want to be kneed. He's a good kid. Really.

"Many accounts imply that we threw all caution to the wind, or that we were heroically courageous. We didn't and we weren't," Iverson said. "We had followed the development of thin-slab casting throughout the decade. We looked before we leaped."[12]

The smaller footprint of Nucor's Crawfordsville factory enabled greater energy efficiency and labor productivity. It took about 3 person-hours to create a ton of sheet steel in an integrated mill, whereas it took 0.6 person-hours in Nucor's facility. In total, the thin-slab technology in smaller facilities enabled 20 to 30 percent cost savings ($50 to $75 a ton).[13] The plant began operating in 1989. By 1991, it was producing 700,000 tons of products. In 1993, it expanded to 1.8 million tons per year.[14] In 1992, Nucor opened another mill in Hickman, Arkansas, and, in 1997, its third mill, in Berkeley Country, South Carolina.

The major integrated steel mills had underestimated Nucor. In 1992, they could no longer ignore it. But they remained unworried. A spokesperson for U.S. Steel told the *Wall Street Journal* that it was ready to build its own minimill by the end of the decade and it would do so with technology that would be even better than Nucor's technology.

That never happened.

Bethlehem's End

Leader after leader invested in the plant that was Bethlehem Steel's namesake. At some point, that plant began to show its age. Former employee Donald Young proclaimed it the "most diverse steel plant there ever was in the whole world."[15] With fifty-six years of industry experience and thirty-three years working at the plant, he would know. He joined Bethlehem Steel in 1962 and called the plant "an operating museum. . . . [I]nstead of phasing this place out, they were investing millions of dollars in this place, scrunched up in an old plant with terrible interplant transportation, so it would never again produce to its capacity."

In 1969, at a groundbreaking ceremony for the Martin Towers a twenty-one-story building that Bethlehem Steel's executive staff moved into in 1972, the company's chair and CEO Edmund Martin (yes, the building's namesake) discussed the centrality of Bethlehem to the company. "We have done a lot of thinking about the probable future of the city of Bethlehem and the Lehigh Valley," he said. "It looks good to us. . . . In short, it is evidence that we are here to stay."[16]

Lewis W. Foy, who followed Martin as CEO, rejected advice to shut down a plant in Johnstown, Pennsylvania (in the western part of the state, about 225 miles from Bethlehem). "I can't shut down mills in my hometown," he said in 1974. "We can make those plants profitable again."[17]

Except they couldn't. Bethlehem Steel bounced around over the next two decades, with good years ($342 million in profits in 1974 and a record $403 million in profits in 1988) and bad years (a net loss of $488 million in 1977 and waves of layoffs and a failed merger with British Steel in 1991). Finally, in 1995, it discontinued steel production in its Bethlehem plant. In 1998, it stopped all industrial activity in the plant. In 2001, the same year I first heard Clay present his story, it filed for Chapter 11 bankruptcy protection. In 2003, it was acquired by International Steel Group. ISG cofounder Wilbur Ross became a billionaire as the conglomerate consolidated and optimized the scraps of the steel industry. In 2019, Martin Tower was demolished.

In the end, the collapse of Bethlehem Steel is a bit of a puzzle. Nucor generally used widely available technologies, albeit ones that were sometimes unproven. It did everything in plain sight of well-capitalized incumbents. Iverson provided detailed information about his costs and his strategy. Yet, when Clay studied the steel industry as part of his research into disruptive innovation, not a single North American integrated producer had invested in minimill technology.

Bethlehem Steel had demonstrated a willingness to make investments—huge investments—throughout its history. It had demonstrated an ability to operate effectively at almost unimaginable scale. But when it came to following a disruptive approach, the company struggled.

Jim Collins compares Nucor and Bethlehem Steel in his 2001 bestseller *Good to Great*. After rattling off a range of statistics, such as how Nucor's profit per employee was ten times that of Bethlehem's, he suggests that Bethlehem struggled because the management team grew fat, stupid, and lazy. "We came to the conclusion that Bethlehem executives saw the very purpose of their activities as the perpetuation of a class system that elevated them to elite status," he writes. At Bethlehem, "people focused their efforts on negotiating the nuances of an intricate social hierarchy, not on customers, competitors, or changes in the external world."[18]

He points, for example, to Martin Tower. With a height of 330 feet, the building was the tallest in the Lehigh Valley. It was shaped like a cross so that there would be more corner offices for executives. Those offices had ornate furniture, doorknobs engraved with the company's logo, and handwoven carpets. The executive cafeteria served delicacies such as filet mignon and lobster tails.

Clay's take is different. "The integrated steel companies' march to the profitable northeast corner of the steel industry," he writes in *The Innovator's Dilemma*, "is a story of aggressive investment, rational decision-making, close attention to the needs of mainstream customers, and record profits."[19]

There's a third possibility.

The Ghosts of Bethlehem

In June 2024, Thomas Wedell-Wedellsborg, a friend and fellow author who reviewed multiple iterations of this book, sent me a photo of Bethlehem Steel's abandoned namesake plant. "It looks utterly awesome," he wrote. Indeed, photos of the abandoned plant (figure 10-2) are haunting.

Figure 10-2

Blast furnace D at Bethlehem Steel's abandoned plant, 2009

Source: Reprinted with permission from Matthew Christopher/Abandoned America.

Why didn't Bethlehem Steel invest in competing minimill technology? Why didn't it acquire a minimill producer in the late 1960s or early 1970s, when these mills were small and affordable? Or even in the early 1980s? It isn't that Bethlehem Steel couldn't have done it. Investing in the new business would not have violated a law of nature. The investment wouldn't have broken any laws. Bethlehem Steel could have. But it didn't. It made a choice.

Clay called that choice rational, but is that the right word for it? Would a rational company make a series of decisions that ultimately led to it shutting down its flagship facility that was the lifeblood of a town for more than a hundred years, declaring bankruptcy, and selling the company to a consolidator who shut down more facilities, laid off more workers, and slashed employee pensions? No one wants that outcome.

Saying Bethlehem's decisions weren't entirely rational benefits from hindsight, of course. A general challenge for disruption is that during a disruptive shift, the right choices look like the wrong ones. And one thing I've learned is that making the right choice is even harder because of the ghosts that haunt every organization.

Ghosts. Powerful but invisible forces that inhibit responding to disruptive change.* Let's look again at Bethlehem Steel. Its rich history had triumph and tragedy, success and struggles. It spawned ghosts that made it even harder to rise to the challenge of disruptive change.

Imprinted Patterns

When ducks are born, they famously follow whatever they see first. Academic research shows that organizations absorb pieces of their founding environment and characteristics of their founders. Behaviors forged during key historical moments end up being remarkably persistent.

Think of the imprint of Bethlehem Steel's glory days in the first half of the twentieth century. Carnegie's formula turned into Grace's maxim: "Always more production." Grace also echoed another Carnegie dictum: "Pioneering doesn't pay." Beyond the Bethlehem beam,

---

* The academic background for the ghost concept was an executive master's program I attended at INSEAD in Singapore from 2019 to 2021. The program was grounded in systems psychodynamics, which focuses on the interaction between collective structures, norms, and practices in social systems and the thoughts, motivations, and emotions of the members of those systems. I think of systems psychodynamics as organizational behavior meets psychiatry. Our program focused on developing the "night vision" to see what is in the shadows. Going through a program on change in the midst of the Covid-19 pandemic was an experience! If you are interested in additional details beyond what's here, you know where to find me.

Bethlehem Steel cautiously put new technologies on hold, instead focusing on operating at scale, keeping its plants running, and sewing up key customers. It hired talented executives and rewarded them handsomely, with Grace himself the second-highest paid executive in the United States in 1941 and 1942, behind Hollywood mogul Louis B. Mayer.

Phrases by iconic leaders become wired into decision-making processes and disappear into "that's just the way we do things around here." Decades of experience taught Bethlehem executives that there will be a Bethlehem moment: World War I, the postwar building boom, World War II, the postwar expansion. But hidden forces can influence and sometimes drive resource allocation decisions in directions that can stymie effective response to disruptive change.

Forging America, a book about Bethlehem Steel and published by Bethlehem's leading newspaper, detailed a visit to Grace's grave in the city's Nisky Hill Cemetery. A nameless writer asked Wilbur Kocher, an electrician at Bethlehem Steel, for directions to the grave. Kocher pointed toward a huge stone memorial just thirty feet away from a cliff with the Lehigh River below and the old Bethlehem plant across the river.

"He's got chairs around it, although nobody ever sits there," Kocher said. "I don't know if it's true, but I heard that they buried him standing up so he could look at the blast furnaces."[20]

Not true. But Grace's ghost played a role in what led to that blast furnace's ultimate shutdown.

Historical Trauma

Human traumas, ranging from wars to intense inter-sibling rivalries, often carry through generations. Traumatic events at companies can create ghosts that lead people to unconsciously shy away from doing things that trigger associations to a painful past. These events include mass firings and exogenous shocks like the Covid-19 pandemic. Other examples include major failures (consider AT&T's $100 billion writedown related to failed acquisitions of DirecTV and Time Warner), abusive leaders, and scandals like accounting misstatements or ethical lapses.

What past trauma haunted Bethlehem Steel? Labor conflict. In the 1960s, steel minimills began gaining traction. One key to their model was to avoid locations that had unionized labor. That's not to say they treated their workers poorly. Nucor had a lean corporate staff, with a couple dozen people in its two-thousand-square-foot headquarters in North Carolina. It placed significant autonomy in the hands of its workers. It primarily motivated workers through generous bonuses based on company and individual productivity. When a new plant opened, about half of its workers would turn over in the first year of operations. Turnover would then drop to single digits as Nucor found people who fit its model, and vice versa.

The Nucor model—grounded in low-cost, cutting-edge technology; based in low-cost environments; and run by efficient labor that worked hard because of the prospects of generous performance bonuses—was powerful.

And on paper, Bethlehem Steel executives could surely see the attractiveness of the Nucor model. But the idea of opening *that* door, of exploring something with nonunionized labor, of setting the labor relations clock back three decades, if not more . . . that could have been just too much.

Fear of Identity Loss

Brené Brown from the University of Houston has produced a powerful set of research and writing about shame and vulnerability. She notes that the biggest driver of shame is the fear of irrelevance. Disruptive change can threaten an organization's very identity. That's scary!

Imagine you were a top executive in Bethlehem Steel in the 1960s. Whom did you work for?

You worked for the company that built America because of your company's expertise in scale operations. Grace described this very quality in 1943:

> *You may recall that last January, I pledged that Bethlehem in 1943 would build a ship a day, a major-size ship a day, with several to spare.... I don't mind saying that there were many times during the year I was doubtful of reaching our goal. And I know that our shipyard managers at times thought that it just couldn't be done.*
>
> *... I believe that Bethlehem was able to handle this enormous shipbuilding job because of the way we are set up as an integrated company. By "integrated" I mean we are a company which produces materials all the way from the ore in the ground to the complete ship. This helps us understand how each part fits together. So I say to you men in the mines in Michigan and Minnesota, in Pennsylvania and in West Virginia that your work is the starting point.*[21]

You worked for a great company, with a pristine image. If you visited its headquarters in the early 1950s, someone like Ellie Prizznick, the daughter of a Bethlehem Steel straightener, might have escorted you. She could have seamlessly swapped with that era's flight attendants. Here's how *Forging America* describes the escorts:

> *Her eyebrows were plucked. Her blond hair was perfectly coiffed. Her face was painted with the same expensive makeup worn by New York models.*
>
> *In the tradition of the pretty young Bethlehem Steel escorts who guided visitors through the company's thirteen-story headquarters on Third Street in South Bethlehem, she stood in front*

of the elevator with the proper posture: one hand under an elbow, one foot in front of another. . . . Bethlehem Steel had an image to uphold.[22]

And you didn't work for U.S. Steel. You didn't work for National Steel. You didn't work for Integrated Steel. You worked for *Bethlehem* Steel.

"The steel that built half of New York City was rolled in those mills and they just let it die," said Tom Jones, the former union president for the United Steelworkers, local union in Bethlehem. "That's just wrong. Have we completely forgotten where we came from?"[23]

Maybe the problem wasn't that Bethlehem Steel forgot where it came from. Maybe the problem was it could never forget its past.

Bethlehem Steel's Echo: Electric Vehicles and Disruption's Extended Shadow

One of the joys of researching and writing this book has been finding hidden points of connection. Julia Child has her first lunch in the town where Joan of Arc was burned at the stake? Nucor's origin story intersects in a winding way with Henry Ford?* Beautiful.

Lehigh Valley Live published a photo history of Martin Tower on May 17, 2019, two days before its implosion. The twenty-third photo and its caption caused me to stop: "Two large mosaics depicting the Gutenberg press and Stephen Daye press are removed from Martin Tower in September 2017 and moved to the National Museum of Industrial History" (figure 10-3).[24]

As I wrote this chapter, that picture stayed with me. It led to me returning to the shadow cast by disruption.

Gutenberg's printing press boomeranged on its backer, the Catholic Church, as the innovation spawned religious revolutions. I'll leave it to

---

* Missed that? It was in a footnote earlier in this chapter.

Figure 10-3

The ghosts of Bethlehem Steel

Source: Photo from Steve Novak, "Martin Tower Demolition: A Photo History of the Bethlehem Steel Headquarters Before Its Implosion," *Lehigh Valley (PA) Live*, May 17, 2019, www.lehighvalleylive.com/news/g66l-2019/05/50278d4ede2239/a-photo-history-the-rise-and-fall-of-martin-tower-before-its-imploded.html. Photo credit: Glenn Koehler, National Museum of Industrial History.

theologians to debate whether those revolutions ultimately strengthened the Catholic Church. At the time, however, it undoubtedly led to pain and struggles. Steel minimills powered disruptive growth that led to the creation of powerful new companies, drove productivity improvements, and lowered prices. But the disruption also led hundreds of thousands of people to lose their jobs and caused towns to be ripped apart. Creative destruction carries a heavy transaction tax.

This chapter shows another layer to that shadow. Also lurking in the dark are forces that make disruption even more difficult than Clay proclaimed in his research. Yes, you need to master the resource allocation process to make sure you don't fall into the trap of optimizing today at the expense of competing effectively tomorrow. You also have to understand that resource allocation decisions are made by humans, who have biases and suffer from blind spots and who, without knowing it, repeat past patterns, shy away from historical traumas, and avoid change that threatens self-identity.

Think about Tesla. Disruption purists look at Tesla with skepticism. Disruption starts with simple solutions at the low end of the market or in new contexts. Tesla started with a premium-priced, sophisticated solution. Yet, it seems to have played out in a disruptive way with market leaders struggling to respond.

There are undoubtedly some disruptive elements to Tesla. Its cars offer new benefits, of course, but because drivers have to wait for the vehicle to charge, they suffer from the dreaded range anxiety. There are some nuances to the story. Tesla, for example, got hundreds of millions of dollars of loans from the US government in 2010, when it was still a young company (it repaid the loan in 2013). And then governments around the world gave ample tax credits to people who bought electric vehicles.

And there are ghosts. From Henry Ford's Model T, the industry has had a dominant approach. Cars are mechanical. Gas powers them. The complexity requires that major upgrades be considered carefully and slowly. The dominant gene is mechanical engineering. An electric-powered vehicle requires understanding software as much as mechanical engineering; indeed, many producers view themselves as software providers who happen to make cars. A line worker will give you all sorts of rational reasons why changing course is challenging, but the unstated reason very well might be the fear of identity loss.

Navigating disruptive change requires understanding that which lurks in the shadow.

11

Anomalies Wanted

The iPhone and the Era of Smart Innovation

"This is a day I have been looking forward to for two and a half years. Every once in a while, a revolutionary product comes along that changes everything . . . Well, today we're introducing three revolutionary products of this class. The first one is a wide-screen iPod with touch controls. The second is a revolutionary mobile phone. And the third is a breakthrough Internet communications device. So. Three things. A wide-screen iPod with touch controls, a revolutionary mobile phone and a breakthrough internet communications device."[1]

You can probably picture the black turtleneck, the round glasses, the jeans. It is January 9, 2007, we're at Macworld, and Steve Jobs is in his element.

"An iPod, a phone and an Internet communicator." The audience is with him. He starts again. "An iPod, a phone . . ." He trails off. He's ready.

"Are you getting it? These are not three separate devices; this is one device, and we are calling it iPhone. Today, Apple is going to reinvent the phone. And here it is."

Behind him appears a picture of Apple's iconic iPod portable music player with a rotary dial.* Laughter. Before finally holding the *real* product onstage, he talks about its purpose. He says traditional mobile phones are relatively easy to use but not very smart, and the smartphones that have been introduced in the past few years are smarter but hard to use.

"We don't want to do either one of these things," Jobs says. "What we want to do is make a leapfrog product that is way smarter than any mobile device has ever been and super easy to use. This is what iPhone is."

What happens next is well known. After the iPhone, Apple launched the iPad, AirPods, and the Apple Watch. It became the world's most valuable company. Jobs died in 2011, but his successor, Tim Cook, broadened Apple's portfolio, and the company increased its value twelvefold in the decade after Cook took the reins. Apple became the first company listed on US stock exchanges to cross $1, $2, and $3 trillion in aggregate value.

What else is there to say about the iPhone? Like any story, the closer you look, the more that interesting facets emerge. As the final case study in the book, the iPhone provides a way to return to the four questions posed in the introduction.

---

* That's another old-fashioned phone quip. Are you ready for this? Phones used to have a round dial with ten holes in them. You would put your finger in the hole representing the digit you wanted to dial and rotate it. Jobs was having fun with the fact that its iPod music player had a scroll wheel. Ever the performer. And it won't be too long before the iPod itself needs to be explained. To younger readers: before the iPhone, Apple had a breakthrough portable music player that reinvented the category and resuscitated a company that, believe it or not, was on the brink of irrelevance if not extinction.

Who Does It?

One of the iPhone's breakthrough features was the ability for users to interact with the screen in new ways. Pinching to zoom in and out, swiping to move to the next picture, double tapping to highlight content all seem mundane today but felt magical when introduced in 2007.

The journey to the iPhone's magical interface started in the 1960s. At the time, computer scientists had begun to experiment with multitouch screen interfaces. E. A. Johnson from the Royal Radar Establishment in Malvern, United Kingdom, created the first functional touch screen in 1965. It worked, but not particularly well, on televisions powered by cathode ray tubes. The first commercialized touch screens trace back to an accidental discovery by G. Samuel Hurst, who was studying atomic physics in 1970. The technology then began appearing in focused applications like automated teller machines.

In the 1980s, the "godfather of multitouch," Bill Buxton, created a multitouch tablet at Xerox's famed Palo Alto Research Center (which Jobs had visited a few years earlier). Buxton's tablet showed the potential to bring the technology to much more mainstream uses. The world wasn't quite ready, however.

In 1999, a young student named Wayne Westerman completed the dissertation for his engineering PhD, "Hand Tracking, Finger Identification, and Chordic Manipulation on a Multi-Touch Surface." He dedicated the dissertation to his mother, Bessie, "who taught herself to fight chronic pain in numerous and clever ways, and taught me to do the same."[2] After back surgery, Westerman's mother developed chronic back pain. Westerman could empathize as he personally was experiencing carpal tunnel syndrome.

In light of his research, he began working on a device that could recognize the difference between single taps and multiple touches and that could translate multiple-finger input on its keyboard into complex commands. He patented the device in 2001 and cofounded a company called FingerWorks with Jeff White, who had worked in Hewlett-Packard's peripheral business.

In 2002, FingerWorks launched a product called the iGesture Pad, targeting a niche market of people dealing with repetitive strain issues. Here is one description of the product:

> *The iGesture Pad was a large touchpad that allowed users to control their computers using hand and finger gestures, replacing the traditional mouse. Imagine you are reading a long document and want to scroll down. Instead of scrolling the mouse wheel or dragging the scrollbar, with the iGesture Pad, you simply place two fingers on the pad and slide them down, as if you're physically pushing the document down. The same gesture can be performed in reverse to scroll up. You can also zoom in or out of a document by placing your thumb and index finger on the pad and sliding them apart or together.*[3]

Sound familiar?

How did this device get to Apple? Maybe it was sheer luck. But luck, as always, is the residue of design. Back in California, Apple had a team working on a secret project to develop a screen that would allow a multitouch interface. Historically, touch screen interfaces like ATMs would have multiple layers of screens that would recognize when the screens were pressed together. Known as *resistive touch screens*, they traced back to Hurst's 1970 discovery and allowed only limited interactivity. The team at Apple was working on *capacitive sensing*, which essentially recognizes the electric charge from a finger.*

---

* A few years ago, I remember seeing a nanoscientist at a health-care conference in New Zealand demonstrate how to use your nose to activate your touch screen if you were wearing gloves in cold weather (the idea made sense at the time). Of course, today there are special gloves that have that ability. God forbid you had to wait to be inside to send that Snap or post that TikTok! Did I get that right, kids?

In his book *The One Device: The Secret History of the iPhone*, Brian Merchant describes how one of the members of the team, Apple engineer Tina Huang, came to work one day with "an unusual, plastic black touchpad marketed to computer users with hand injuries."[4] It was Westerman's device.

Maybe Huang's use of the device created a serendipitous connection. Maybe it was publicity from the product winning an award from the Consumer Electronics Show in 2005. Whatever the path, Apple acquired FingerWorks later that year. To the chagrin of its loyal users, the company discontinued the iGesture Pad. To the joy of billions of smartphone users around the world, the FingerWorks technology played a critical part in making the iPhone so brilliantly easy to use.

. . .

Innovation is not the job of the few; it is the responsibility of the many. Is innovation an individual activity or a collective one? Yes. The lone inventor is a myth. Even the book's first story where we can name the protagonist—Johannes Gutenberg and his printing press—wasn't a solo story. Gutenberg had business partners, a financial backer (who sued him; such is life), customers, and so on. Yes, there has to be a visionary who sees what others don't.* But it is simply impossible for anything to be achieved by an individual. Innovation is a collectively individualistic act. You don't have to be a hero to play a role in disruptive change. Anyone can, and everyone should, play a part in doing something different that creates value.

Buxton believes that the development of multitouch illustrates what he calls the long nose of innovation: "The bulk of innovation behind

---

* That line is from the song "For the Dreamers." That song, which is from the musical *Back to the Future*, was on a Spotify playlist I listened to most frequently while writing this book. It's worth a listen!

the latest 'wow' moment . . . is also low-amplitude and takes place over a long period—but well before the 'new' idea has become generally known, much less reached the tipping point. It is what I call The Long Nose of Innovation."⁵ Disruption rarely traces to a single big-bang invention moment. Rather, everything builds over time. *The Sources of Invention*, one of the first methodical studies of the history of big ideas, puts it well: "Every inventor, however original he may appear to have been, is laying bricks upon a building which has long been in the course of construction from innumerable and mainly unknown hands."⁶

Is It Random?

Despite the significant hype, the iPhone was not an immediate breakthrough success. Of course, some of Apple's most devoted fans waited to be the first to get their hands on the device, with Apple saying it sold 270,000 units in thirty hours. But the perfectly imperfect product had limitations. There were only sixteen apps on the phone. It wasn't possible to add more. Heck, it wasn't even possible to customize the wallpaper. It was available only through one carrier in the United States.* It ran on slow networks. The battery life was short. And it was priced at a significant premium.

Jobs was famously a fan of tight control. After the failed launch of a phone called ROKR with Motorola, he noted that "we're not very good at going through orifices to get to the end user."**⁷ And if he saw

---

* That carrier was Cingular, which would become AT&T Wireless. While the company bears the name AT&T, it has almost no connection to the company originally founded by Alexander Graham Bell. *Seeing What's Next*, by Clay, me, and Erik Roth, discusses the history of wireless telephony in greater depth.

** The ROKR would be a fun tracer. The phone launched in 2005. The cover of *Wired*'s October issue said it all: "You Call This the Phone of the Future?" The device was returned at six times the industry average. So, no.

the potential of a vibrant ecosystem of developers at launch, it wasn't obvious. "What's the killer app?" Jobs said. "The killer app is making calls. It's amazing how hard it is to make calls on most phones."[8]

But that killer app wasn't killer enough. "The iPhone was almost a failure when it first launched," said Brett Bilbrey, who ran Apple's media architecture group. "Many people don't realize this. Internally, I saw the volume sales—when the iPhone was launched its sales were dismal. It was considered expensive; it was not catching on. For three to six months . . . it was not doing well."[9]

Why? It was simple, Bilbrey said. "There were no apps."

. . .

Innovation is predictably unpredictable. There are clear patterns that cut across the stories in this book. Magic happens at intersections. The first market for a disruptive idea is one that finds the solution perfectly imperfect. The special sauce of disruption is a business model that allows the unique creation, delivery, and capture of value. All disruptions feature a version of the hero's journey. While the broad patterns are predictable, every journey has its unique surprises and struggles. You can forecast disruption. You can't plan it.

It was predictable that the first version of the iPhone was imperfect. It was predictable that even the great Steve Jobs had it wrong. What exactly Apple would do to improve the phone was unpredictable. That it would work hard to make the device better, to wrinkle out imperfections, to seek to expand its market was perfectly predictable.

When evaluating a disruption, remember that what you see in a given moment is an artifact of past decisions and there's likely something coming next. And that's the question. What's next? When you make decisions on a snapshot of a moment, you will miss the full impact of a disruption. The full movie *always* has plot twists as companies find innovative ways to expand their reach.

If you were Nokia, Research in Motion, Motorola, Palm, Handspring, or Ericsson, maybe you took solace in the iPhone's slow start. And then, in October 2007, Jobs changed course. It shouldn't have really been a surprise. When the iPod first came out, it was limited because it only seamlessly worked with Apple's Mac computers. When Apple ported iTunes to Windows-based computers, the product took off. Jobs had demonstrated that he was willing to be flexibly controlling.

Jobs announced that Apple would have a software development kit for third-party apps by February 2008. "We are excited about creating a vibrant third-party developer community around the iPhone," he wrote on Apple's Hot News site, "and creating a vibrant third-party developer community around the iPhone and enabling hundreds of applications for our users."[10]

Apple pioneered a model where it would take a 30 percent cut of all the apps sold on its platform. That was a small but material number when the app ecosystem was emerging and Apple had ten million devices in the market. With a developed ecosystem and more than one billion active iPhones, even with a drop in commissions to 15 percent, the App Store generated close to $100 billion in revenues for Apple in 2023. By way of comparison, the Coca-Cola Company earns approximately $50 billion a year in revenue. Think about that the next time you start playing a silly game on your phone.

Is It Accelerating?

Days before the iPhone officially went on sale, *Wall Street Journal* reporters Walt Mossberg—in some circles as iconic as Jobs—and Katherine "Katie" Boehret offered their thoughts on the product. The mostly positive review drew special attention to the screen:

> The iPhone is simply beautiful. It is thinner than the skinny Samsung BlackJack, yet almost its entire surface is covered by a huge,

> vivid 3.5-inch display. . . . The display is made of a sturdy glass, not plastic, and while it did pick up smudges, it didn't acquire a single scratch, even though it was tossed into Walt's pocket or briefcase, or Katie's purse, without any protective case or holster. No scratches appeared on the rest of the body either.[11]

The touch screen took us to the 1960s. Understanding the screen's beauty requires going back a decade earlier to visit an iconic innovator in Upstate New York. The company Corning was founded in 1851 by Amory Houghton as the Bay State Glass Company. It took on its modern name when it moved to Corning, New York, in 1868 under Houghton's son. The company specializes in glass, ceramics, and optical materials. Its first big breakthrough was developing the glass for Thomas Edison's light bulb.

In the early 1950s, Donald Stookey was experimenting with photosensitive glass. He put lithium silicate into a furnace to bake. He set the temperature to 600°C, but a malfunctioning controller made the temperature shoot up to 900°C.

Stookey assumed he would open the furnace door to a disaster scene. Instead, he found an off-white plate that, when it slipped out of the tongs he used to extract it from the furnace, bounced. Stookey had discovered the world's first synthetic glass-ceramic, a material that could tolerate high heats without expanding. Corning incorporated the material, Pyroceram, into a line of dishes and bowls under the brand CorningWare (Corning divested its consumer businesses in 1998).

Corning then started a development effort called Project Muscle to make CorningWare clear. Scientists discovered that adding aluminum oxide to a glass composition and then dousing the glass in a bath of hot potassium salt did the trick.

Corning called the product Chemcor. It was strong. So strong it could withstand 100,000 pounds of pressure per square inch (psi), or about fourteen times more pressure than normal glass could withstand. You could see through it. But Corning couldn't find a viable commercial application. It tried car windshields, but it was *too* strong. As *Wired* reporter Bryan Gardiner noted in his extensive history of Chemcor, "Crash tests found that 'head deceleration was significantly higher' on the windshields—the Chemcor might remain intact, but human skulls would not."[12]

In 1971, Corning shut the product down after spending $42 million in development (about $300 million today).

More than three decades later, Jobs decided that the iPhone should have a glass screen instead of the original plastic design. And Jobs being Jobs, he made this decision months before launch. Fortunately, in 2005, the success of Razr V3, a stylish flip phone launched by Motorola, had led Corning to dust off Chemcor to see if the product could work for cell phones and watches. Corning named the new (old) effort Gorilla Glass. Success was possible, scientists thought, if chemists could make the old Chemcor samples thinner.

When Jobs called in early 2007 to ask about the possibility of a 1.3-millimeter, ultrastrong glass, Corning developers weren't sure they could make even a *sample* of the product. They weren't sure they could manufacture it. They weren't sure they could manufacture it at scale. And they certainly weren't sure they could do all this in six months.

Jobs had a memorable meeting with Wendell Weeks, a Corning lifer who had become its CEO in 2005. As Weeks described Gorilla Glass, Jobs was skeptical that it would meet Apple's specifications and "started explaining to Weeks how glass was made." Weeks, no shrinking violet, responded with, "Can you shut up and let me teach you some science?" Jobs was sold, but Weeks said delivering the quantities Jobs was seeking wasn't possible. Jobs's biographer Walter Isaacson describes what happened next:

> *"Don't be afraid," Jobs replied. This stunned Weeks, who was good-humored and confident but not used to Jobs's reality distortion field. He tried to explain that a false sense of confidence*

would not overcome engineering challenges, but that was a premise that Jobs had repeatedly shown he didn't accept. He stared at Weeks unblinking. "Yes, you can do it," he said. "Get your mind around it. You can do it."

As Weeks retold this story, he shook his head in astonishment. "We did it in under six months," he said. "We produced a glass that had never been made." Corning's facility in Harrodsburg, Kentucky, which had been making LCD displays, was converted almost overnight to make Gorilla Glass full-time. "We put our best scientists and engineers on it, and we just made it work." In his airy office, Weeks has just one framed memento on display. It's a message Jobs sent the day the iPhone came out: "We couldn't have done it without you."[13]

A modest $20 million line of business became a booming $700 million business in just a few years. By 2024, an estimated *eight billion* mobile devices had used Gorilla Glass.[14] Here's how Gardiner described it in 2012:

> In just five years, Gorilla Glass has gone from a material to an aesthetic—a seamless partition that separates our physical selves form the digital incarnations we carry in our pockets. We touch the outer layer and the body closes the circuit between an electrode beneath the screen and its neighbor, transforming motion into data. It's now featured on more than 750 products and 33 brands worldwide, including notebooks, tablets, smartphones, and TVs. If you regularly touch, swipe, or caress a gadget, chances are you've interacted with Gorilla.[15]

It wasn't just five years, of course. There was a five-*decade* gap between Stookey's accidental discovery of a synthetic glass-ceramic and the moment Jobs's famed reality distortion field aided his effort to have the iPhone's screen made of Gorilla Glass.

. . .

Innovation requires patient perseverance. There is a lot of talk about the accelerating pace of change, how, for example, ChatGPT was the fastest innovation in history to get to one hundred million users. Yes, that specific *use* of AI was adopted quickly. However, ChatGPT launched in what was arguably year sixty-six for AI, if you trace it back to a 1956 conference at Dartmouth College hosted by Professors John McCarthy and Marvin Minsky. What looks like rapid adoption in a short period is a kink in a curve that has been moving slowly for decades.* The first six years after Ray Kroc took license to McDonald's, the company's aggregate profits were less than $200,000. The transistor achieved its intent of displacing vacuum tubes in telecommunications networks only after decades of development. Julia Child's call to adventure came in 1951, but she weathered two rejections before her landmark cookbook launched in 1961. Pampers failed three separate test markets. The disciplined experimentation that resulted from the work of Sir Francis Bacon and like-minded thinkers provides the technical keys to managing this journey. There's also an emotional component. Amazon founder Jeff Bezos described how he was willing to be misunderstood for long periods. Disruptive innovation requires being comfortable . . . with being uncomfortable.

---

* Horace Dediu, a technology analyst who graduated from Harvard Business School and worked at the Clayton Christensen Institute for Disruptive Change for a few years, did extensive analysis about technology adoption curves. He presented his ideas to Innosight, and my TL;DR takeaway was that yes, when an innovation modularly layers into an existing infrastructure—think Instagram or ChatGPT—adoption can be very fast. If an innovation requires a completely new infrastructure—think electric vehicles or the metaverse—it still takes decades. A sixty-minute presentation goes through the argument in more depth (see Department of Information and Communications Engineering at Aalto University, "Horace Dediu, Clayton Christensen Institute, the Modularity Revolution: How Markets Are Created," presentation at Telecom Forum 2015, YouTube video, https://www.youtube.com/watch?v=Mk4Ov7WXm_w).

Is It a Universal Good?

Pick up the iPhone. Turn it over in your hand. Think about the materials that went into it. A teardown of an iPhone 6 (circa 2016) device found elements you would probably expect, such as aluminum, iron, lithium, and silicon, and elements you might not, such as gold, magnesium, and tantalum. Those materials don't magically appear. Author Brian Merchant visited mines in Bolivia, where the life expectancy of full-time miners was just forty years. "Your iPhone," Merchant writes, "begins with thousands of miners working in often brutal conditions on nearly every continent to dredge up the war elements that make its components possible."[16]

And, of course, those materials must become a finished product. For years there was a stamp on the back of the iPhone: Designed in California by Apple, Assembled in China. Very likely in an enormous facility in Shenzhen owned by the Taiwanese Hon Hai Precision Industry Company, known more commonly as Foxconn. Despite Apple's famed secrecy, Merchant managed to sneak into the heart of Foxconn's Longhua facility in 2016:*

> *It is block after block of looming, multiple-story, gray, grime-coated cubes. It is factories all the way down, a million consumer electronics being threaded together in identically drab monoliths. You feel tiny among them, like a brief spit of organic matter between aircraft carrier-size engines of industry. It's factories as far as you can see; there is simply nothing beautiful in sight.*
>
> *In fact, the only things designed to be aesthetically pleasing, designed to appeal to humans at all, are the corporate mascots*

---

* He did so by saying he needed to use the bathroom. In the end of his very accessible book, he writes, "This is potentially a crime of trespassing in China, but I feel it is justified due to the history of abuse and tight media controls by Foxconn. It seemed to me it would be a public service to get a fair and un-spun image of the factory."

and the trimmed hedges back near the food court, and that feels grim out here—in Longhua, you're either in a strip mall or on the factory floor.[17]

It's Ford's assembly line perfected, with workers performing more than a hundred steps to produce a phone in twenty-four hours. One worker, in charge of wiping a special polish on the display, said she worked on seventeen hundred devices in a day. If the worker did a twelve-hour shift, that is a phone every twenty-five seconds, not factoring in breaks. In 2010, there were reports of eighteen deaths by suicide at the facility, leading then CEO Terry Gou to install nets to catch falling bodies.[18]

I'm not picking on Apple. I'm sure it works hard to make its supply chain as ethical as possible, and its production processes as safe as possible. That work is hard. The world is complex.

And, of course, there's what people actually do with the phone. In 2024, Jonathan Haidt's book *The Anxious Generation* drew on research that showed a sharp rise in depression and anxiety in teenagers, particularly teenage girls, after 2012, a point at which smartphones like the iPhone and social media apps like Instagram became widely available in developed markets. The scientific community debated Haidt's conclusions, but few disagreed that phones have an alluring, almost addictive quality to them. There is no doubt that putting massive computing power in your pocket provides massive benefits. But there's another side to the story.

. . .

Disruption casts a shadow. Today, innovation is viewed as a universal good. It wasn't always. As described earlier, in 1548, Kind Edward VI issued a proclamation Against Those That Doeth Innouate. Innovation wasn't a heroic act; it was an act of heresy. Disruption challenges the status quo, which might be good for the world but less good for those seeking to preserve that status quo. Stare at the abandoned Bethlehem Mill and feel the ghosts that haunted it, making it legitimately and understandably hard for an iconic company to grapple with the disruptive power of the steel minimills. The introduction of disruption can be messy. Did city-living pedestrians hail the arrival of Ford's Model T? Of course not. They protested. The 1920s featured a battle between those berating jaywalkers and those shaking their fists at flivverboobs. We know who won. Disruption changes the dynamics in organizations, industries, and society. That's good for some and bad for others. If you don't see the shadow that disruption casts, it can swallow you.

The iPhone's Echo: Modernizing Disruption Theory

The Apple iPhone allows us to return full circle to disruptive theory itself. Clay had his doubts about the iPhone at its launch:

> *The iPhone is a sustaining technology relative to Nokia. In other words, Apple is leaping ahead on the sustaining curve [by building a better phone]. But the prediction of the theory would be that Apple won't succeed with the iPhone. They've launched an innovation that the existing players in the industry are heavily motivated to beat: It's not [truly] disruptive. History speaks pretty loudly on that, that the probability of success is going to be limited."*[19]

Whoops.

Clay famously had a sign in his office that read "Anomalies Wanted" (figure 11-1).* As part of the research for this book, I caught up with my old friend Clark Gilbert to trade memories about Clay. Gilbert's experience gives him a unique way to frame Clay's work and its impact. Gilbert began his career as a consultant for Monitor Group, the consulting company cofounded by Harvard Business School legend Michael Porter. He was a doctoral student at HBS in the late 1990s. His award-winning thesis was on how incumbents responded to disruptive change. He taught for several years in HBS's entrepreneurial management department before running a media company and a university. He was an adviser to Innosight during this period. In 2019, he assumed a role with the Church of Jesus Christ of Latter-day Saints (of which Clay was also a member), overseeing its education assets.

Gilbert specifically reflected on how Joe Bower pushed his doctoral students to investigate anomalies. "It's definitely in all of our research profiles, but Clay was the first one to codify it," said Gilbert, who coauthored a book chapter titled "Anomaly Seeking Research" with Clay. "He was the master of that. Clay would say find an industry where the best theories would predict *A*, and you are seeing *B*."[20]

Clay thought the iPhone would fail. It obviously didn't. That's an anomaly. Good. Studying it leads to three important lessons about using disruptive innovation theory.

First, think about the comparison set. Clay later said he mistakenly compared the iPhone to other *phones*, where he really should have been comparing it to *laptop computers*. Indeed, Mossberg and Boehret titled their June 2007 review, "The iPhone Is a Breakthrough Handheld Computer." When you make that comparison, the disruptive

---

* More about that sign. Clay loved woodworking. His eldest son, Matt, described how when he was young, his father carved a set of bowling pins that Matt and his four siblings could attempt to knock down with a softball. A few years later, Matt got a woodburning tool that . . . largely stayed in his closet. One year, his father found it and gave each of the children bowling pins with a hand-inscribed message. And then, Clay created the sign. At one point, this book was titled *Anomalies Wanted*.

Figure 11-1

Sign hanging in Clay Christensen's office

Source: Reprinted with permission from Katie Zandbergen.

nature of the iPhone comes into focus. The iPhone certainly had its limitations. Its battery didn't last long. Typing wasn't easy. The first version had only a handful of applications. It had new advantages, such as a cutting-edge internet browser and the kind of radical simplicity that had been Apple's hallmark for decades. In classic disruptive fashion, the iPhone brought computing to new contexts, where people were delighted with a somewhat limited product.

Second, expand your view. The iPhone wasn't just a *product* play by Apple; it was an *ecosystem* play. Ron Adner, from the Tuck School of Business at Dartmouth College, has deeply studied ecosystems in his excellent books *The Wide Lens* and *Winning the Right Game*. His research shows how innovation success requires smart thinking about how to work with suppliers, channels to market, complementary industries, and more. Analyzing disruption has grown harder as the world has gotten more complicated. This chapter has touched on several pieces of the iPhone but could have picked Apple's porting its Macintosh operating system onto a handheld device, the uniqueness of its web browser, the power of its camera, its custom-created semiconductor, its unique partnership with Google, and on and on. It is increasingly hard to say a company is or isn't following a disruptive path, because a company's product or service combines so many different facets. Zoom in and look at the pieces, and then zoom out to look at the broader ecosystem.

Third, focus on the movie, not the snapshot. The full disruptive potential of the iPhone wasn't readily apparent when it launched. The launch of the App Store and the explosion of third-party apps turned what could have been an interesting niche product into a full-powered disruption, one that would accelerate as more mobile phone carriers in more countries began to offer the iPhone and as Apple continued to layer more technological capabilities into the phone. The best forecasters regularly update their forecast as new data arrives. When it comes to disruption, expect the unexpected and adjust accordingly.

The first of these lessons surfaces another apparent anomaly. If the iPhone was disruptive to the personal computer, disruptive theory would predict that Apple—a leading personal computing manufacturer—would not allocate resources to it. Disruptive theory would similarly predict that Meta, the parent of Facebook, would underinvest in photo apps and messaging services. But the company acquired and reasonably smoothly integrated Instagram and WhatsApp, developing an integrated portfolio of social media offerings. Disruptive theory would also predict that once Microsoft stagnated in the early 2000s, it would never recover. Instead, under the leadership of Satya Nadella, Microsoft made major, early investments in cloud technology and AI, trading places with Apple and Nvidia as the world's most valuable company throughout the early part of the 2020s.

There is no law of the universe that says incumbents *can't* respond to disruptive change. Yes, of course, people will point to various constraints, such as regulators, shareholders, and boards of directors, and those are real, but they aren't of the level of ice freezes at 32°F or gravity causes objects to accelerate at 9.8 meters per second squared.*

In his biography of Steve Jobs, Isaacson notes how "Jobs was deeply influenced by [Clay Christensen's] book *The Innovator's Dilemma*."[21] If you read and study the forces of disruption, you can turn Clay's original dilemma into an opportunity.

Was Clay upset about being wrong about the iPhone? Not at all. Someone seeking to *prove* their idea would view an anomaly as a failure. That wasn't Clay. He always sought to *improve* his ideas. So, he welcomed anomalies.

The academic literature is rich with discussion and debate about Clay's original research. I'm not an academic, so I'm not going to opine on the discussion, other than to say it is robust. In 2016, Professor Joshua Gans did a wonderful job synthesizing it in an accessible book titled *The Disruption Dilemma*. I had a chance to facilitate a discussion with Gans at the St. Gallen Symposium in Switzerland in 2017. Gans described how he shared this book with Clay, who didn't view it as an attack but instead viewed it as an opportunity to learn. One thing Clay always regretted was his unsuccessful attempt to connect with Harvard historian Jill Lepore when she was working on a *New Yorker* article titled "What the Gospel of Innovation Gets Wrong."

Apple led to results that were similar to historical disruptions but didn't quite fit Clay's original work. That's OK. Bacon and Boyle are happy because learning about the anomalies leads to richer, more robust models. And Clay is happy because anomalies are the best place to learn.

---

* I didn't look up the physics fact. Well, I looked it up to make sure I had it right, but that has been burned in my brain since 1991.

Conclusion

What's Next?

The Future of Disruption

We've seen disruption on the battlefield, in boardrooms, and in the baking of baguettes. We've learned lessons from disruptive doers and disruptive thinkers. The endings of the stories in *Epic Disruptions* are written. We know who won, who lost, who made good decisions, and who made bad ones.

The forces of disruption, however, very much carry on. In June 2022, I was in Dublin, attending a gathering of top leaders in the consumer goods industry. One of the speakers was Alan Jope, the CEO of Unilever. "I must say I am a bit fed up of living in unprecedented times," he said. "I'd like to live in precedented times for a couple of years."[1] The audience laughed.

On September 26, 2022, Unilever announced that Jope planned to retire at the end of 2023. Jope's successor lasted all of 20 months before getting booted out.

There's no going back. Wave after wave of disruption promises continual changes to the global economy. Historians can look back and draw lines between before and after, but living in the middle is messy. What would Clay say, I wondered, if you asked him to make sense of AI, autonomous vehicles, and more?*

There are three things I believe Clay would do. As a consummate storyteller, of course, he would start with a story. I think he'd probably use one of his classics, like the one he liked to tell about Andy Grove (the version that follows is from his speech to the Harvard Business School's graduating class of 2010).**

> *Before I published* The Innovator's Dilemma, *I got a call from Andrew Grove, then the chairman of Intel. He had read one of my early papers about disruptive technology, and he asked if I could talk to his direct reports and explain my research and what it implied for Intel. Excited, I flew to Silicon Valley and showed up at the appointed time, only to have Grove say, "Look, stuff has happened. We have only 10 minutes for you. Tell us what your model of disruption means for Intel." I said that I couldn't—that I needed a full 30 minutes to explain the model, because only with it as context would any comments about Intel make sense. Ten minutes into my explanation, Grove interrupted: "Look, I've got your model. Just tell us what it means for Intel."*
>
> *I insisted that I needed 10 more minutes to describe how the process of disruption had worked its way through a very different industry, steel, so that he and his team could understand how*

---

* This is the third version of the conclusion. The first version was speculative fiction, imaging how disruptive forces would impact my family in 2040. My kids *loved* the fact that they made the book in a material way. Three reviewers I trust said, "It just doesn't work." The next attempt was a chapter-length imagined conversation with Clay. That also didn't quite work. If you want to see either version, just email me at scott.d.anthony@tuck.dartmouth.edu.

** I have probably heard this story more than twenty times. Sometimes the story features Andy Grove; sometimes it features the US Defense Department. It always features steel minimills. Clay truly loved telling the steel minimill story.

disruption worked. I told the story of how Nucor and other steel minimills had begun by attacking the lowest end of the market—steel reinforcing bars, or rebar—and later moved up toward the high end, undercutting the traditional steel mills.

When I finished the minimill story, Grove said, "OK, I get it. What it means for Intel is . . .," and then went on to articulate what would become the company's strategy for going to the bottom of the market to launch the Celeron processor.

I've thought about that a million times since. If I had been suckered into telling Andy Grove what he should think about the microprocessor business, I'd have been killed. But instead of telling him what to think, I taught him how to think—and then he reached what I felt was the correct decision on his own.

That experience had a profound influence on me. When people ask what I think they should do, I rarely answer their question directly. Instead, I run the question aloud through one of my models. I'll describe how the process in the model worked its way through an industry quite different from their own. And then, more often than not, they'll say, "OK, I get it." And they'll answer their own question with more insight than I could have.[2]

Give people a model, namely, disruptive innovation. Provide them with a meaningful metaphor. And let them do the work.

Next, Clay would push for using a common language. He would often say how one of his biggest regrets was choosing the word *disruption* to describe the phenomenon he observed in his doctoral research.

"I just didn't realize how that would create such a wide misapplication of the word 'disruption' into things that I never meant it to be applied to," he once said.[3]

The basic dictionary definition of *disruption* refers to "a disturbance" or "a radical change." Clay's research focused on an innovation that made the complicated simple and the expensive affordable, disrupting and redefining the basis of competition in a market. As we have seen throughout this book, a Christensen-style disruption has massive growth potential that historically was captured by an entrant, not the market-leading incumbent.

It is a critical distinction. A dictionary-style disruption presents its challenges, but if it doesn't compete in new ways, it can generally be handled by existing structures and systems. A Christensen-style disruption, on the other hand, requires active management and careful organizational design. That expends leadership energy that could be used for other purposes. That's why it is important to make sure something truly requires that kind of active management; otherwise, that energy could best be put to other uses.

Clay had a simple way to help people think through this issue: "When the founder of a company tells me they are disruptive, the first question I always ask is, 'To what?' This is an important question, because disruption is a relative concept."[4] Indeed, you can understand how disruptive something is only if you have a basis for comparison.

Finally, I'm sure Clay would push for questions rather than answers. I remember distinctly a discussion I had with Clay in 2007 when I was considering a doctoral degree. His advice was to make sure I had a great question on my mind. That way, I would look at everything through the lens of that question; doing so would help me make connections and speed up my thesis research and writing.

"I've long believed that asking the right questions is the only way to get to the right answer," Clay once said. "And understanding what questions to ask takes real work."[5]

So, let's go through ten ongoing disruptions that connect to stories in this book. I list the disruptions in alphabetical order and, for each one, pose three questions to consider based on the stories in this book. Consistent with Clay's guidance, there are no projections, no guesses as to winners and losers, and no stock picks. That's all up to you.

Additive Manufacturing (Pampers)

The Pampers story was about low-priced convenience. The key to delivering that convenience was developing a manufacturing process that allowed reliable production of diapers at high volumes and low prices. If you go to a diaper manufacturing facility, you see large machines moving at astonishing speeds.

There are numerous categories of manufacturing processes. A job shop. An assembly line. A continuous process. A batch process. And, in 1987, a new option joined the list: *additive* manufacturing. In 1987, 3D Systems, a company formed by an inventor named Charles Hull, introduced a sophisticated printer that used a technique called stereolithography to solidify plastic with a laser, essentially building something from the ground up, layer by layer. Additive manufacturing contrasts with subtractive manufacturing, where the essence of process is removing materials. Cutting. Sawing. Lathing. Milling. All are forms of subtractive manufacturing.

To date, additive manufacturing has largely been used by hobbyists or to produce highly specialized parts that aren't economically produced at scale. It has some clearly disruptive characteristics, such as introducing new performance trade-offs, bringing production to less centralized locations, and enabling the creation of new ecosystems. As it continues to improve, additive manufacturing has the potential to dramatically reshape any industry that involves making or distributing things.

Questions to consider

1. The disruptive giants of the internet era succeeded because of their disruptive business models. What new business models can additive manufacturers create to strengthen their disruptive potential?

2. One way health-care disrupters have thrived is by enabling lesser-trained providers to do more. What would it look like if the hobbyist ecosystem pulled in new applications, such as home improvement?

3. What technological development would be analogous to thin-slab casting, which allowed steel minimills to advance into higher market tiers?

Artificial Intelligence (Printing Press)

Johannes Gutenberg brought together "the arts and the adventures" to create the printing press. As the disruptive innovation spread through Europe and beyond, it also spread discussion, dissent, and revolution. Newspapers, radio, television, and the internet allowed information to spread, sometimes for better, sometimes for worse.

On November 30, 2022, OpenAI released ChatGPT, and the world seemed to be on the edge of another revolution. The ability of generative artificial intelligence (gen AI) to instantly generate fluid prose, create evocative images, and produce impossible-to-detect fake videos held untold disruptive potential. Consultants and academics seemed in a race to one-up one another with proclamations about how the nature of work would change.

The overnight success was anything but. As described earlier, the term *artificial intelligence* was coined at a 1956 conference at Dartmouth College hosted by John McCarthy and Marvin Minsky. That conference featured the presentation of a proof-of-concept AI program called the Logic Theorist, which mimicked human problem-solving. Research split into subfields such as machine learning, gen AI, and natural language processing.

Fast-forward to 2025. If you were to go to a conference on innovation, you would surely see a video of a glorious world with AI. Your day was better organized, your meals had perfect nutritional balance, your skin glistened. If you want a contrasting view, you could flip on the TV series *Black Mirror*, where AI controls, contains, distracts, distorts.

The foundational nature of AI means it has the potential to have disruptive impact on scores of industries, similar in kind to the waves of disruption unleashed by the collection of technologies that made the internet available to the layperson in the 1990s.

Questions to consider

1. Just as Gutenberg mixed "the arts and the adventures," how might humans mix with machines to drive disruption in unique ways?

2. Is there an equivalent to the McDonald's real estate model that will enable AI-powered business models to drive disruption?

3. What ghosts will market-leading incumbents making big investments in AI need to overcome?

Autonomous Vehicles (Model T)

Autonomous vehicles began with a blind man. When he was just five years old, Ralph Teetor accidentally blinded an eye while playing with a knife. A year later, he lost sight in the other eye from the trauma. The trope that losing one sense strengthens another was perhaps true for Teetor, who developed a strong ability to develop complex visualizations. He earned an engineering degree from the University of Pennsylvania and, with some relatives, formed a company called Perfect Circle, eventually becoming its president. In 1950, he earned a patent for the Speedostat, a way to regulate the speed of a vehicle without human intervention. Today we call this technology *cruise control*.

Many modern vehicles steer, accelerate, and brake autonomously, with humans ready to intervene when required. Increased levels of autonomy have humans playing a decreasing role, from monitoring a driverless car remotely to having no connection to the car at all.*

The automobile is so central to the world that a transition to autonomous driving could have knock-on impacts on industries well beyond the manufacturing and distribution of automobiles. For example, will there still be auto insurance in a world of full autonomy? Most likely. Anytime there is a risk of something bad happening, there is an opportunity to make a market in that risk. What is likely to change is who takes out the insurance policy. It might not be you or me; it might be the organizations creating the algorithms that power autonomous cars. Stretch your imagination further, and you can see the possibility of radically redesigned cities with space formerly used for parking transformed into different uses.

Disruptive potential galore!

Questions to consider

1. Gunpowder was first used as a tool for magicians. What use cases will welcome a solution that still has limitations instead of viewing those limitations as showstoppers?

2. The automobile's development accelerated after the jaywalkers defeated the flivverboobs. What is the equivalent for autonomous vehicles?

3. What industries beyond insurance and urban planning could be affected by autonomous vehicles?

---

* The Society of Automotive Engineers has six levels of autonomy, from zero (no automation) to five (full automation).

Cleantech (Gunpowder)

People learned how to harness the explosive power that resulted from mixing sulfur, saltpeter, and charcoal. Over the centuries, they also learned how to harness electrical power, the power trapped in fossil fuels, and then the power contained within atoms. These developments gave rise to the modern world but also created a challenge, as greenhouse gas emissions threaten to raise temperatures and potentially even destabilize the earth.

In *How to Avoid a Climate Disaster*, Bill Gates notes how getting to carbon neutral required addressing the fifty-two billion tons of carbon dioxide equivalents that power today's modern world. His view is that there is no technological silver bullet. Perhaps. Innovators around the world are working furiously to prove him wrong. Small modular reactors that enable clean nuclear plants that can be run safely and efficiently. Carbon dioxide removal technologies, some of which essentially suck carbon out of the air. Ever-advancing solar and wind technologies. Are we reaching a tipping point that will dramatically reshape the $6 trillion global energy market?

Questions to consider

1. What is the hearing aid equivalent for next-generation technologies like direct air capture?

2. What emergent business models might allow tomorrow's disrupters to thrive in the new reality?

3. What ghosts will leading energy companies need to overcome to grab hold of the disruptive potential of new technologies?

Distributed Ledger Technologies (Julia Child)

Julia Child was concerned about sharing her recipes before they were published, because she viewed them as proprietary and didn't want them to be stolen. The use of distributed ledger technologies has the potential to reshape how creators distribute and make money from their creations.

The technology traces back to a 2008 paper that framed the idea for an electronics payment system where two people could transact directly with each other without needing an intermediary like a bank.[6] In 2009, the first mechanism based on the idea—Bitcoin—became available to the public. The next fifteen years featured waves of growth and unexpected developments.

Distributed ledger technology—in essence a database that everyone has access to and can therefore use to see the history of ownership of an asset or changes to a contract—has created a class of speculators, a surge and retreat in non-fungible tokens showing unique ownership of digital assets, and experiments with smart contracts.

Questions to consider

1. Part of Child's impact was her ability to empathetically connect with her customers. What truly unique problems—beyond money laundering and other nefarious uses—can be solved by distributed ledger technologies?

2. The Catholic Church served as a critical customer that helped boost the printing press. Is there a similarly transformational customer for digital ledger technologies?

3. How else could these technologies enable the creation of new business models in the maker space?

Drones (the Transistor)

Some observers suggest that World War II interrupted the work William Shockley, John Brattain, and Walter Bardeen were doing on the transistor. My view is that the war accelerated the transistor's development by providing the trio with diversifying experience and creating new military applications that, like radar, pushed the boundaries of scientific knowledge. It also dramatically increased the government's willingness to invest in technology.

Similarly, drones have gone from largely being toys for hobbyists to must-haves for militaries. In 2017, twenty-four-year-old inventor Palmer Luckey cofounded Anduril Industries, a defense technology company specializing in advanced autonomous systems. The war in Ukraine served as the coming-of-age moment for military drones. Couple military developments and company experiments with drone-based delivery, and the disruptive potential of drones becomes ever clearer.

Questions to consider

1. Drones trace back to work at the Defense Advanced Research Projects Agency.* Are there other DARPA developments that could enable drones to enter new consumer applications?

2. What is the hearing aid equivalent for these applications?

3. A recurrent theme in this book is that innovation magic happens at intersections. What magic could occur at the intersection of additive manufacturing and drone-based delivery?

---

* There was a whole ode to DARPA cut from the transistor chapter. Beyond drones, DARPA helped spur global positioning technology, carbon composite materials, stealth technology, and Siri. See "Special Forces' Innovation: How DARPA Attacks Problems," a great 2013 *Harvard Business Review* article by two DARPA leaders.

Mixed Reality (the iPhone)

Apple's iPhone was arguably history's most successful product. Apple sold about 250 million phones in 2023 and cleared more than $1.5 trillion in iPhone sales between 2014 and 2023. That's a hard act to follow.

Apple launched its TV in 2007 (the same year it introduced the iPhone), the iPad in 2010, the Apple Watch in 2015, AirPods in 2016, and its HomePod smart speakers in 2018. And then, in 2024, it launched Vision Pro, its augmented/virtual-reality goggles. The first version of the product cost $3,500, a steep premium over Meta's $699 Quest 3. That makes Apple a premium-priced, reasonably late entrant into a small but material category. Sound familiar?

Virtual reality involves full immersion in a simulated environment. Augmented reality involves blending the physical and the virtual. Mixed reality covers both use cases. Imagine a car dashboard that overlays information, or a smart contact lens that allows information to appear at the edge of your vision. Yes, the means of accessing virtual and augmented reality available in 2025 are big and clunky, and they subject some users to motion sickness, but they are improvements over Google's failed Google Glass experiment from the 2010s. Over the next decade, goggles, smart contacts, implants, or whatever gets invented will surely be significantly smaller, cheaper, faster, and better.

Questions to consider

1. The iPhone was a masterful concoction of ingredients, many of which were decades in the making. That's innovation's so-called long nose. What seeds that were planted between 2000 and 2019 will bear their innovation fruit in the next decade?

2. Ultimate success for Apple required the creation of a new ecosystem: the device, plus the mobile phone carrier, plus the

third-party App Store. What different components need to come together for mixed reality to become a . . . reality?

3. What are potential second- or third-order effects of scaled mixed reality?

New Food (McDonald's)

Cows. They are an essential ingredient for McDonald's. In 2040, will we say they *were* an essential ingredient for McDonald's? The cows that are processed into McDonald's hamburgers today are a borderline ecological disaster for several reasons. First, cows are an inefficient way to provide protein, because they produce less protein than the protein required to feed them. Second, cows take up a significant amount of arable land. About half of US land area is agricultural. Third, cows produce methane, which is an even more nefarious greenhouse gas than carbon dioxide because methane lingers longer in the atmosphere.

Disruption to the rescue! *New food* is a blanket term to cover a range of emerging technologies. Precision fermentation allows the laboratory-based synthesis of molecules that occur in food products like milk and yeast. Cell-based meat is grown in a bioreactor but is otherwise identical to meat from animals. Plant-based meat, an alternative to meat, is made entirely from plants.[7]

Like any potentially disruptive development, new food has taken its bumps. Leading plant-based meat provider Beyond Meat saw its valuation plunge by more than 95 percent between 2020 and 2024. The company's rival Impossible Foods shelved plans to go public, because of wavering demand. Lab-grown meat providers struggled to figure out how to economically produce their product; the first lab-grown hamburger allegedly cost more than $250,000, and even massive price declines still left it five times more expensive than traditional beef.

Remember that technology develops faster than people's lives change to adapt to the technology. That which is not good enough and that which is too expensive today inevitably gets better and gets cheaper. The snapshot of the moment can distract us from the movie and its predictable ending.

Questions to consider

1. The "special sauce" at McDonald's was a business model that allowed it to create, deliver, and capture value. What disruptive business model will emerge from new food? Could it be centralized production or perhaps something like a molecule design service?

2. One key to McDonald's success was how closely it worked with its suppliers. As described earlier in the book, it spent years perfecting its fries. Who will partner with new-food providers to develop something that is tasty and affordable?

3. Existing food processors like Tyson Foods and Cargill have invested in new food. Have they studied disruption theory enough to successfully manage a potential industry transition?

Robotics (Steel)

You know for sure that if longtime Nucor CEO Ken Iverson were alive today, he would be all over robotics. He demonstrated a consistent willingness to push the edge, always experimenting with new technologies that improved efficiencies and allowed him to sell steel at ever-lower prices.

Collaborative robots, otherwise known as cobots, emerged in the 1950s. You can probably picture hydraulic arms that can pick up and manipulate things that are too heavy for humans to easily lift. Those arms got stronger, got faster, got more mechanized.

Fears began to mount that the robot revolution would eradicate manufacturing jobs. A 2013 study by Oxford exacerbated those fears, with headlines blaring that 47 percent of jobs were at risk (that's not what the study said, but nuance and headlines don't often go well together).[8] Around that time, Amazon began a string of acquisitions to build up a robotic infrastructure for its warehouses and distribution centers. The Covid-19 pandemic drove further interest in industrial robotics. And research studies began to show that companies that invested in robotics ended up *adding* employment.[9]

And, of course, there are those Boston Dynamics robots . . . the dog-like Spot on display at the Boston Science Museum both transfixes and terrifies my children.

Questions to consider

1. Companies have generally used robots to improve the efficiency of current manufacturing and distribution systems. What minimill equivalent of a rethought model grounded in robots could have sustainably more attractive economics than traditional players?

2. Nucor disrupted the integrated steel mills that were haunted by the ghost of Andrew Carnegie's mantra of "run the mills full." What ghosts will make it hard for existing incumbents to respond to companies that base their business on robotics?

3. What disruptive potential exists at the intersection of AI and robotics? How about at the intersection of robotics and smart health?

Smart Health (Florence Nightingale)

Disruption in health care involves three interrelated elements: lesser-trained providers doing what once required specialized expertise, shifts in the point of care from centralized to decentralized locations, and movements from treatment to prevention. The development of step-by-step rules enables disruption. Florence Nightingale's remarkable visuals drove a societal shift toward prevention, and her work to build schools and write manuals enabled broader populations to serve as effective nurses.

Smart health is a blanket term for a set of solutions that have the potential to radically reconfigure health care. It includes several advancements:

- *Scientific advances* introducing new therapies and techniques for treating disease, such as the dramatic decreases in the cost of sequencing the human genome; technologies like CRISPR, which allow the targeted editing of genes; the emergence of new technologies like m-RNA drugs and vaccines; digital therapeutics for difficult-to-treat conditions like addiction; and additive manufacturing, which could significantly transform certain surgical procedures.

- *Digital health solutions* that transform how consumers monitor and manage their health. Advancements include wearable devices, remote monitoring, and remote appointments.

- *Enabling technologies* like AI and advanced analytics that drive value and performance in the health-care space through improved diagnostics, better population health insights, faster pharmaceutical development, and better care management.

Questions to consider

1. Is there a Crimean War equivalent that galvanizes public interest in smart health? Who could be a modern-day Lady with the Lamp?

2. Florence Nightingale used data and charts to change the discourse on health. What parallel development could dramatically advance smart health?

3. Simplifying nursing into basic language and do-it-yourself principles enabled a broader population to care for patients. How can smart health produce smart *outcomes* in ways that drive easy adoption and use?

Final Thoughts

One of Clay's common quips was that his wife, Christine, would say that he had the disruptive line diagram etched onto his glasses. Everywhere he looked, he saw disruption. He taught me how to see the world through a similar set of lenses. It let me see what I would otherwise miss.

The stories in this book show how innovation has gone from something that threatened the king to something that is celebrated but can still be threatening. Watch out for ghosts! The stories show how innovation is more collaborative than ever before, but they still have heroes. There's no *I* in *team*, but there is one in *innovation*. You see how the patterns of innovation are clearer, but there are still detours and surprises on the road to success. Innovation is predictably unpredictable, and overnight success often takes decades. So be prepared to patiently persevere!

I'm an optimist by nature, which I suspect comes across in my writing. I believe the disruptive changes unfolding today will make the world a better place. I'm not immune to the reality that, for example, a significant amount of cryptocurrency use has been by bad people doing bad things. Or that many new media technologies get traction in what you might politely call less savory applications. Or that the ability to precisely manufacture food can lead producers to make bad things even more addictive than they are. Disruption casts a shadow, and there are many, many issues to confront. The most challenging ones are likely those we can't see today. Nonetheless, disruption has historically allowed more people to do things that matter to them. It transforms what exists and creates what doesn't, by making the expensive affordable and the complex simple. I believe in the power of disruption to make the world a better place.

The world is a complicated place, and there are no easy answers. Clay's teaching transformed how I looked at the world. My own research, fieldwork, and teaching have further refined what I see and how I make sense of the world around me. I hope this book similarly sharpens your lenses, helping you see what you would otherwise miss. What you do with that, of course, is up to you.

Good luck.

Notes

Introduction

1. Tom Nicholas, "How History Shaped the Innovator's Dilemma," *Business History Review* 95 (spring 2021): 121–148.
2. Kim Clark, interview with author, December 4, 2023.
3. Clayton M. Christensen, "The Rigid Disk Drive Industry: A History of Commercial and Technological Turbulence," *Business History Review* 67 (winter 1993): 531–588.
4. Herbert E. Casson, *The History of the Telephone* (Chicago: A.C. McClurg & Co., 1910), accessed from Project Gutenberg, https://www.gutenberg.org/files/819/819-h/819-h.htm.

Chapter 1

1. From William Shakespeare, *Othello*, act 3, scene 3.
2. Francis Bacon, *Novum Organum* (1620), p. 86.
3. Charles River Editors, *Gunpowder: The History and Legacy of the Explosive That Modernized Warfare* (Charles River Editors, 2021), 3. Hereafter cited as *Gunpowder: The History*.
4. *Gunpowder: The History*, 3.
5. *Gunpowder: The History*, 5.
6. *Gunpowder: The History*, 5–6.
7. Jack Kelly, *Gunpowder: Alchemy, Bombards & Pyrotechnics; The History of the Explosive That Changed the World* (New York: Basic Books, 2005), 7.
8. *Gunpowder: The History*, 16.
9. *Gunpowder: The History*, 9.
10. Kelly, *Gunpowder*, 31.
11. Kelly, *Gunpowder*, 47.
12. Kelly, *Gunpowder*, 47.
13. *Gunpowder: The History*, 30.
14. Kelly, *Gunpowder*, 50.

15. Kelly, *Gunpowder*, 53.

16. Chip Heath and Dan Heath, *Switch: How to Change Things When Change Is Hard* (New York: Broadway Books, 2010), 15.

17. INSEAD, "The Decline of Nokia: Interview with Former CEO Olli-Pekka Kallasvuo," interview by Quy Huy, YouTube video, March 12, 2014, https://www.youtube.com/watch?v=jR5a_DBYSmI.

18. Richard S. Tedlow, *Giants of Enterprise: Seven Business Innovators and the Empires They Built* (New York: Collins, 2003), 424.

19. Kelly, *Gunpowder*, 33. Emphasis added.

20. *Gunpowder: The History*, 11.

Chapter 2

1. Stephan Füssel, *Gutenberg*, translated by Peter Lewis (London: Haus Publishing, 2019), 2.

2. John Man, *The Gutenberg Revolution: The Story of a Genius and an Invention That Changed the World* (London: Bantam Books, 2009), 46.

3. Füssel, *Gutenberg*, 7.

4. Man, *Gutenberg Revolution*, 57.

5. Man, *Gutenberg Revolution*, 61. Emphasis in original.

6. Füssel, *Gutenberg*, 19.

7. Füssel, *Gutenberg*, 26.

8. Man, *Gutenberg Revolution*, 130–133.

9. Füssel, *Gutenberg*, 30.

10. Füssel, *Gutenberg*, 41.

11. Füssel, *Gutenberg*, 43.

12. Man, *Gutenberg Revolution*, 189.

13. Füssel, *Gutenberg*, 152.

14. Man, *Gutenberg Revolution*, 14.

15. Füssel, *Gutenberg*, 22.

16. Steven Johnson, *Where Good Ideas Come From: The Natural History of Innovation* (New York: Riverhead Books, 2010), 152.

17. Man, *Gutenberg Revolution*, 264.

18. Claudia H. Deutsch, "At Kodak, Some Old Things Are New Again," *New York Times*, May 2, 2008.

Chapter 3

1. Benoît Godin, *Innovation Contested: The Idea of Innovation Over the Centuries* (New York: Routledge, 2015), 2.

2. Giovanni Aquilecchia, "Giordano Bruno: Italian Philosopher," in *Encyclopaedia Britannica Online*, last updated January 1, 2025, https://www.britannica.com/biography/Giordano-Bruno.

3. Godin, *Innovation Contested*, 281.

4. Francis Bacon, quoted in Daphne Du Maurier, *The Winding Stair: Francis Bacon, His Rise and Fall* (Garden City, NY: Doubleday, 1976), 195.
5. Du Maurier, *Winding Stair*, 20.
6. Francis Bacon, quoted in Du Maurier, *Winding Stair*, 24.
7. Newberry Library, "Title Page of Novum Organum," Newberry Library Miscellaneous Files, uploaded September 5, 2020, https://archive.org/details/nby_231406.
8. Francis Bacon, quoted in Steven Shapin, *The Scientific Revolution* (Chicago: University of Chicago Press, 2018), 20.
9. Francis Bacon, quoted in Du Maurier, *Winding Stair*, 144.
10. Shapin, *Scientific Revolution*, 20.
11. Shapin, *Scientific Revolution*, 20.
12. Plato, quoted in Shapin, *Scientific Revolution*, 58.
13. Francis Bacon, quoted in Shapin, *Scientific Revolution*, 139.
14. Francis Bacon, quoted in Shapin, *Scientific Revolution*, 66.
15. Francis Bacon, quoted in Shapin, *Scientific Revolution*, 87.
16. Francis Bacon, quoted in Shapin, *Scientific Revolution*, 88.
17. David C. Lindberg, *The Beginnings of Western Science: The European Scientific Tradition in Philosophical, Religious, and Institutional Context, Prehistory to A.D. 1450* (Chicago, University of Chicago Press, 2007), 51.
18. Du Maurier, *Winding Stair*, 143.
19. King James, quoted in Du Maurier, *Winding Stair*, 143.
20. Shapin, *Scientific Revolution*, 96.
21. Shapin, *Scientific Revolution*, 98.
22. Robert Boyle, *New Experiments Physico-mechanicall Touching the Spring of Air and Its Effects* (Oxford: printed by H. Hall, Printer to the University, for Tho. Robinson, 1660), 106, available at https://archive.org/details/chepfl-lipr-AXA74/page/n9/mode/2up?q=%22principal+fruit%22//.
23. Godin, *Innovation Contested*, 19.
24. Godin, *Innovation Contested*, 19.
25. Godin, *Innovation Contested*, 215.
26. Robert Robinson, quoted in Godin, *Innovation Contested*, 143–144.
27. Godin, *Innovation Contested*, 282.

Chapter 4

1. Michael Friendly, "The Golden Age of Statistical Graphics," *Statistical Science* 23, no. 4 (November 2008): 502–535.
2. Hourly History, *Florence Nightingale: A Life from Beginning to End* (CreateSpace Independent Publishing Platform, 2018), 14.
3. Alfred, Lord Tennyson, "The Charge of the Light Brigade," Poetry Foundation, uploaded 2017, https://www.poetryfoundation.org/poems/45319/the-charge-of-the-light-brigade.
4. R. J. Andrews, ed., *Florence Nightingale: Mortality and Health Diagrams* (San Francisco: Visionary Press, 2022), 19.

5. Adam Rodman, host, "The Lady with the Lamp," *Bedside Rounds*, podcast, episode 42, January 13, 2019, http://bedside-rounds.org/episode-42-the-lady-with-the-lamp/.

6. Andrews, *Florence Nightingale: Mortality and Health Diagrams*, 30.

7. John Macdonald, quoted in UK National Archives, "Florence Nightingale: Why Do We Remember Her?," National Archives, Classroom Resources, https://www.nationalarchives.gov.uk/education/resources/florence-nightingale/.

8. Andrews, *Florence Nightingale: Mortality and Health Diagrams*, 30.

9. Andrews, *Florence Nightingale: Mortality and Health Diagrams*, 36.

10. Andrews, *Florence Nightingale: Mortality and Health Diagrams*, 38.

11. Andrews, *Florence Nightingale: Mortality and Health Diagrams*, 60.

12. Andrews, *Florence Nightingale: Mortality and Health Diagrams*, 56.

13. Andrews, *Florence Nightingale: Mortality and Health Diagrams*, 68.

14. Andrews, *Florence Nightingale: Mortality and Health Diagrams*, 16.

15. Florence Nightingale, *A Contribution to the Sanitary History of the British Army During the Late War with Russia* (London: John W. Parker and Son, 1859).

16. Florence Nightingale, *Mortality of the British Army, At Home and Abroad, and during the Russian War, as Compared with the Mortality of the Civil Population in England* (London: Harrison and Sons, St. Martin's Lane, 1858), 6–7.

17. Friendly, "Golden Age of Statistical Graphics."

18. Andrews, *Florence Nightingale: Mortality and Health Diagrams*, 95, 108.

19. R. J. Andrews, quoted in Lilly Smith, "The Untold Story of How Florence Nightingale Used Data Viz to Save Lives," *Fast Company*, May 18, 2021, https://www.fastcompany.com/90637011/the-untold-story-of-how-florence-nightingale-used-data-viz-to-save-lives.

20. Adam Grant (@AdamMGrant), "The pen may not be mightier than the sword, but its ink lasts longer," Twitter (now X), October 19, 2018, https://x.com/AdamMGrant/status/1053309222220369921.

21. Virginia Dunbar, foreword to *Notes on Nursing: What It Is, and What It Is Not*, by Florence Nightingale (New York: Dover Publications, 1969; republication of first American edition, published by D. Appleton and Company in 1860), xi–xii.

22. Nightingale, *Notes on Nursing*, 127.

23. Nightingale, *Notes on Nursing*, 125.

24. Nightingale, *Notes on Nursing*, 88.

25. The primary source for this section was Kevin Hazzard, "Freedom House Ambulance Service," *99% Invisible*, podcast, episode 405, July 7, 2020, https://99percentinvisible.org/episode/freedom-house-ambulance-service/.

26. Committee on Trauma and Committee on Shock, Division of Medical Sciences, National Academy of Sciences, "Accidental Death and Disability: The Neglected Disease of Modern Society," National Research Council, Washington, DC, September 1966.

27. Committee on Trauma and Committee on Shock, "Accidental Death and Disability," 12.

28. Hazzard, "Freedom House Ambulance Service."

29. Hazzard, "Freedom House Ambulance Service."

Chapter 5

1. Jeffrey L. Rodengen, *The Legend of Nucor Corporation* (Ft. Lauderdale, FL: Write Stuff Enterprises, 1997), 12.

2. "The Modern Moloch," *99% Invisible*, podcast, episode 76, April 4, 2013, https://99percentinvisible.org/episode/episode-76-the-modern-moloch/, accessed February 13, 2025.

3. Peter D. Norton, *Fighting Traffic: The Dawn of the Motor Age in the American City* (Cambridge, MA: MIT Press, 2011), 210.

4. Allan Nevins, with the collaboration of Frank Ernest Hill, *Ford: The Times, the Man, the Company* (New York: Scribner's, 1954), 54, quoted in Richard S. Tedlow, *Giants of Enterprise: Seven Business Innovators and the Empires They Built* (New York: Collins, 2003), 143.

5. Donn Werling, quoted in Russ Banham, *The Ford Century: Ford Motor Company and the Innovations That Shaped the World* (New York: Artisan, 2002), 22.

6. Ford Bryan, quoted in Banham, *Ford Century*, 20.

7. Nevins and Hill, *Ford: The Times, the Man, the Company*, 152, quoted in Tedlow, *Giants of Enterprise*, 147.

8. Banham, *Ford Century*, 28.

9. Tedlow, *Giants of Enterprise*, 126.

10. Tedlow, *Giants of Enterprise*, 151.

11. Tedlow, *Giants of Enterprise*, 119.

12. Tedlow, *Giants of Enterprise*, 154.

13. Banham, *Ford Century*, 86.

14. Bob Casey, quoted in Banham, *Ford Century*, 193.

15. Joseph S. Seaton, "The Small Car and Its Uses," *Harper's Weekly* 54, no. 2768 (January 1, 1910), 26, https://archive.org/details/sim_harpers-weekly_harpers-weekly_1910-01-08_54_2768/page/26.

16. Henry Ford, *My Life and Work* (Garden City, NY: Doubleday, 1923).

17. Henry Ford, quoted in Banham, *Ford Century*, 99.

18. Banham, *Ford Century*, 42.

19. Banham, *Ford Century*, 242.

20. Theodore Levitt, "Marketing Myopia," *Harvard Business Review*, July–August 2004.

21. Ray Kroc, with Robert Anderson, *Grinding It Out: The Making of McDonald's* (Chicago: H. Regnery, 1977), 30–31.

22. Ford R. Bryan, quoted in Banham, *Ford Century*, 63.

23. Tedlow, *Giants of Enterprise*, 177.

24. Tedlow, *Giants of Enterprise*, 132.

25. Tedlow, *Giants of Enterprise*, 137.

26. Tedlow, *Giants of Enterprise*, 429.

Chapter 6

1. Michael Riordan and Lillian Hoddeson, *Crystal Fire: The Birth of the Information Age* (New York: Norton, 1997), 1.

2. Riordan and Hoddeson, *Crystal Fire*, 44.

3. Riordan and Hoddeson, *Crystal Fire*, 85.
4. Riordan and Hoddeson, *Crystal Fire*, 127.
5. Riordan and Hoddeson, *Crystal Fire*, 137.
6. Riordan and Hoddeson, *Crystal Fire*, 2.
7. Riordan and Hoddeson, *Crystal Fire*, 145.
8. Riordan and Hoddeson, *Crystal Fire*, 4.
9. Ralph Brown, quoted in Riordan and Hoddeson, *Crystal Fire*, 164.
10. "The News of Radio," page 46 of the *New York Times*, July 1, 1948. Accessed via the New York Times archive, https://timesmachine.nytimes.com/timesmachine/1948/07/01/94947389.html?pageNumber=46.
11. Jack Morton, quoted in Riordan and Hoddeson, *Crystal Fire*, 169.
12. Riordan and Hoddeson, *Crystal Fire*, 7.
13. Pat Haggerty, quoted in Riordan and Hoddeson, *Crystal Fire*, 211.
14. Riordan and Hoddeson, *Crystal Fire*, 217.
15. William Shockley, quoted in Riordan and Hoddeson, *Crystal Fire*, 225.
16. Robert Noyce, quoted in Riordan and Hoddeson, *Crystal Fire*, 237.
17. Gordon Moore, quoted in Riordan and Hoddeson, *Crystal Fire*, 241.
18. Walter Brattain, quoted in Riordan and Hoddeson, *Crystal Fire*, 279.
19. Riordan and Hoddeson, *Crystal Fire*, 275.
20. Steve Jobs quotes by Walter Isaacson, "The Real Leadership Lessons of Steve Jobs," *Harvard Business Review*, April 2012.
21. Nick Hughes and Susie Lonie, "M-PESA: Mobile Money for the 'Unbanked' Turning Cellphones into 24-Hour Tellers in Kenya," *Innovations: Technology, Governance, Globalization* 2, no. 1–2 (2007): 63–81.
22. Matthew J. Eyring, Mark W. Johnson, and Hari Nair, "New Business Models in Emerging Markets," *Harvard Business Review*, January–February 2011.

Chapter 7

1. Julia Child, with Alex Prud'homme, *My Life in France* (New York: Alfred A. Knopf, 2006), 16.
2. Child, *My Life in France*, 18.
3. Bob Spitz, *Dearie: The Remarkable Life of Julia Child* (New York: Alfred A. Knopf, 2012), 3.
4. Spitz, *Dearie*, 39.
5. Spitz, *Dearie*, 49.
6. Spitz, *Dearie*, 154.
7. Child, *My Life in France*, 5.
8. Child, *My Life in France*, 61.
9. Child, *My Life in France*, 116.
10. Child, *My Life in France*, 207.
11. Dorothy de Santillana, quoted in Child, *My Life in France*, 209.
12. Child, *My Life in France*, 215–217.
13. Child, *My Life in France*, 217.
14. Judy Jones, quoted in Child, *My Life in France*, 222.
15. Child, *My Life in France*, 223.

16. Craig Claiborne, "Cookbook Review: Glorious Recipes," *New York Times*, October 18, 1961, https://www.nytimes.com/1961/10/18/archives/cookbook-review-glorious-recipes-art-of-french-cooking-does-not.html.

17. Spitz, *Dearie,* 321.

18. Child, *My Life in France*, 240.

19. Time Staff, "Food: Everyone's in the Kitchen," *Time*, November 25, 1966, https://time.com/4230699/food-everyones-in-the-kitchen/.

20. Child, *My Life in France*, 242.

21. Spitz, *Dearie*, 344.

22. Julia Child, quoted in Time Staff, "Food: Everyone's in the Kitchen."

23. Gill Malinsky, "57% of Gen Zers want to be influencers—but 'it's constant Monday through Sunday,' says creator," CNBC.com, September 14, 2024, https://www.cnbc.com/2024/09/14/more-than-half-of-gen-z-want-to-be-influencers-but-its-constant.html.

24. Child, *My Life in France*, 254.

25. Child, *My Life in France*, 255–256.

26. Spitz, *Dearie*, 383.

27. Child, *My Life in France*, 297. Emphasis in original.

Chapter 8

1. Ray Kroc, with Robert Anderson, *Grinding It Out: The Making of McDonald's* (New York: St. Martin's Paperbacks, 1987), 166, 171, and 201.

2. Kroc, *Grinding It Out*, 8.

3. John F. Love, *McDonald's: Beyond the Arches* (Bantam, rev. ed., 1995), 13.

4. Love, *McDonald's Beyond the Arches*, 21.

5. Kroc, *Grinding It Out*, 10–11.

6. Love, *McDonald's: Behind the Arches*, 141.

7. Fred Turner, quoted in Love, *McDonald's: Behind the Arches*, 114.

8. Love, *McDonald's: Behind the Arches*, 226.

9. Love, *McDonald's: Behind the Arches*, 229.

10. Kroc, *Grinding It Out*, 171–172.

11. Kroc, *Grinding It Out*, 120.

12. Love, *McDonald's: Behind the Arches*, 155.

13. Kroc, *Grinding It Out*, 88.

14. Love, *McDonald's: Behind the Arches*, 113–114.

Chapter 9

1. Procter & Gamble, "Pampers: The Birth of P&G's First 10-Billion-Dollar Brand," P&G UK web page, June 26, 2012, https://www.pg.co.uk/blogs/pampers-birth-pgs-first-10-billion-dollar-brand/.

2. Davis Dyer, Frederick Dalzell, and Rowena Olegario, *Rising Tide: Lessons from 165 Years of Brand Building at Procter & Gamble* (Boston: Harvard Business School Press, 2004), 24.

3. Dyer et al., *Rising Tide*, 68.

4. Bob Duncan, quoted in Harry Tecklenburg, "A Dogged Dedication to Learning," *Research-Technology Management* 33, no. 4 (1990): 12–15.

5. Tecklenburg, "Dogged Dedication to Learning."

6. Tecklenburg, "Dogged Dedication to Learning."

7. John Shiffert, quoted in "Products: The Great Diaper Battle," *Time*, January 24, 1969, https://content.time.com/time/subscriber/article/0,33009,900601-1,00.html.

8. Dyer et al., *Rising Tide*, 134.

9. "Products: The Great Diaper Battle."

10. Dyer et al., *Rising Tide*, 234.

11. Jill Boughton, interview with author, December 20, 2023.

12. Clayton M. Christensen, Taddy Hall, Karen Dillon, and David S. Duncan, *Competing against Luck: The Story of Innovation and Customer Choice* (New York: HarperBusiness, 2016), 88.

13. Andrew Sawyer, "PG: Big 2H product news discussed by A.G. Lafley is……,," Goldman Sachs in email forwarded to author, July 18, 2008.

14. Boughton, interview with author, December 20, 2023.

15. John R. Graham, Campbell R. Harvey, and Shiva Rajgopal, "Value Destruction and Financial Reporting Decisions," *Social Science Research Network*, December 2005, https://papers.ssrn.com/sol3/papers.cfm?abstract_id=871215.

16. Andy Jassy, "Our Origins," video interview with Andy Jassy, Amazon Web Services, https://aws.amazon.com/about-aws/our-origins/.

17. Charles A. O'Reilly III and Michael L. Tushman, *Lead and Disrupt: How to Solve the Innovator's Dilemma* (Stanford, CA: Stanford Business Books, 2016), 47.

18. John Cook, "Jeff Bezos on Innovation: Amazon 'Willing to Be Misunderstood for Long Periods of Time,'" *GeekWire*, June 7, 2011, https://www.geekwire.com/2011/amazons-bezos-innovation/.

19. Henry Mintzberg and James A. Waters, "Of Strategies, Deliberate and Emergent," *Strategic Management Journal* 6 (1985): 257–272.

Chapter 10

1. Ken Iverson, with Tom Varian, *Plain Talk: Lessons from a Business Maverick* (New York: John Wiley & Sons, 1998), 127.

2. Richard S. Tedlow, *Giants of Enterprise: Seven Business Innovators and the Empires They Built* (New York: Collins, 2003), 58.

3. Ardith Hilliard and David Venditta, eds., *Forging America: The Story of Bethlehem Steel*, 2nd ed. (Lehigh Valley, PA: Morning Call, 2010), 39.

4. "Schwab Dies at 77 in His Home Here: Steel Master for Half Century Suffered Heart Attack in August while in London," *New York Times*, September 19, 1939.

5. Hilliard and Venditta, *Forging America*, 60.

6. Hilliard and Venditta, *Forging America*, 45.

7. Hilliard and Venditta, *Forging America*, 92.

8. Hilliard and Venditta, *Forging America*, 110.

9. Ken Iverson, quoted in James C. Collins, *Good to Great: Why Some Companies Make the Leap—and Others Don't* (New York: HarperBusiness, 2001), 76.

10. Jeffrey L. Rodengen, *The Legend of Nucor Corporation* (Ft. Lauderdale, FL: Write Stuff Enterprises, 1997), 95.

11. Rodengen, *Legend of Nucor Corporation*, 110.

12. Iverson and Varian, *Plain Talk*, 150.

13. Iverson and Varian, *Plain Talk*, 154.

14. Rodengen, *Legend of Nucor Corporation*, 113.

15. Jacob Roth, "Bethlehem Steel: The Rise and Fall of an Industrial Titan," *Pennsylvania History: A Journal of Mid-Atlantic Studies* 87, no. 2 (spring 2020): 390–402.

16. Hilliard and Venditta, *Forging America*, 119.

17. Hilliard and Venditta, *Forging America*, 126.

18. Collins, *Good to Great*, 138.

19. Clayton M. Christensen, *The Innovator's Dilemma: When New Technologies Cause Great Firms to Fail* (Boston: Harvard Business Review Press, 2024; reprint of 1997 edition, with a new foreword), 123.

20. Hilliard and Venditta, *Forging America*, 59.

21. Eugene Grace, quoted in Hilliard and Venditta, *Forging America*, 100.

22. Hilliard and Venditta, *Forging America*, 103.

23. Tom Jones, quoted in Hilliard and Venditta, *Forging America*, 136.

24. Steve Novak, "Martin Tower Demolition: A Photo History of the Bethlehem Steel Headquarters Before Its Implosion," *Lehigh Valley (PA) Live*, May 17, 2019, https://www.lehighvalleylive.com/news/g66l-2019/05/50278d4ede2239/a-photo-history-the-rise-and-fall-of-martin-tower-before-its-imploded.html.

Chapter 11

1. Steve Jobs, "Steve Jobs Introducing the iPhone at MacWorld," January 9, 2007, YouTube video posted by superapple4ever on December 2, 2010, https://www.youtube.com/watch?v=x7qPAY9JqE4.

2. Wayne Westerman, "Hand Tracking, Finger Identification, and Chordic Manipulation on a Multi-Touch Surface" (PhD diss., University of Delaware, 1999), https://resenv.media.mit.edu/classarchive/MAS965/readings/Fingerwork.pdf.

3. Arturo Moreno, "FingerWorks: Changing the World with a Touch," *Medium*, June 3, 2023, https://arturojmoreno.medium.com/fingerworks-changing-the-world-with-a-touch-147efa0ab435.

4. Brian Merchant, *The One Device: The Secret History of the iPhone* (New York: Little, Brown, 2017), 21.

5. Bill Buxton, "The Long Nose of Innovation," *BusinessWeek*, January 2, 2008 (revised May 30, 2014), https://billbuxton.com/BW%20Assets/01%20The%20Long%20Nose%20of%20Innovation%20Revised.pdf.

6. John Jewkes, David Sawers, and Richard Stillerman, *The Sources of Invention*, 2nd ed. (New York: W. W. Norton, 1969), 27.

7. Frank Rose, "Battle for the Soul of the MP3 Phone," *Wired*, November 1, 2005. https://www.wired.com/2005/11/phone-2/.

8. Jobs, "Introducing the iPhone."

9. Merchant, *The One Device*, 166.

10. Steve Jobs, quoted in Michael Lafferty, "iPhone SDK (Software Development Kit) Announced," Apply Community web page, October 18, 2007, https://discussions.apple.com/thread/1183763?sortBy=best.

11. Walter S. Mossberg and Katherine Boehret, "The iPhone Is a Breakthrough Handheld Computer," *Wall Street Journal*, June 26, 2007, https://allthingsd.com/20070626/the-iphone-is-breakthrough-handheld-computer/.

12. Bryan Gardiner, "Glass Works: How Corning Created the Ultrathin, Ultrastrong Material of the Future," *Wired*, September 24, 2012, https://www.wired.com/2012/09/ff-corning-gorilla-glass/.

13. Walter Isaacson, *Steve Jobs* (New York: Simon & Schuster, 2011), 471.

14. Ryan Raffaelli, "The Three Traps That Stymie Reinvention," *Sloan Management Review*, August 19, 2024, https://sloanreview.mit.edu/article/the-three-traps-that-stymie-reinvention/.

15. Gardiner, "Glass Works."

16. Merchant, *The One Device*, 53–54.

17. Merchant, *The One Device*, 280.

18. Merchant, *The One Device*, 265.

19. Jena McGregor, "Clayton Christensen's Innovation Brain," *BusinessWeek*, July 1, 2007.

20. Clark Gilbert, interview with author, December 15, 2023.

21. Isaacson, *Steve Jobs*, 432.

Conclusion

1. Alan Jope, quoted in Reuters, "Unilever CEO Sees Crises as New Normal for Industry," *Euronews*, June 22, 2022, https://www.euronews.com/next/2022/06/22/europe-food-conference-unilever.

2. Clayton M. Christensen, "How Will You Measure Your Life?," *Harvard Business Review*, July–August 2010, https://hbr.org/2010/07/how-will-you-measure-your-life.

3. David Skok, "The Man Who Saw Tomorrow's Disruption—and Gave Me Hope for Journalism," *Atlantic*, January 31, 2020, https://www.theatlantic.com/ideas/archive/2020/01/man-who-saw-tomorrows-disruption/605824/.

4. Clayton M. Christensen, "Disruption 2020: An Interview with Clayton M. Christensen," interview by Karen Dillon, *MIT Sloan Management Review*, February 4, 2020, https://sloanreview.mit.edu/article/an-interview-with-clayton-m-christensen/.

5. Christensen, "Disruption 2020."

6. Satoshi Nakamoto (pseudonym), "Bitcoin: A Peer-to-Peer Electronic Cash System," *Social Science Research Network*, August 21, 2008, https://ssrn.com/abstract=3440802.

7. "Precision Fermentation and the Dairy Disruption," RethinkX, 2025, https://learn.rethinkx.com/precision-fermentation/dairy.

8. Carl Benedikt Frey and Michael Osborne, "The Future of Employment: How Susceptible Are Jobs to Computerisation?" Working Paper, Oxford Martin School, September 1, 2013, https://www.oxfordmartin.ox.ac.uk/publications/the-future-of-employment.

9. "Economists Are Revising Their Views on Robots and Jobs," *Economist*, January 22, 2022, https://www.economist.com/finance-and-economics/2022/01/22/economists-are-revising-their-views-on-robots-and-jobs.

Index

Aachen Pilgrimage, 33–34
"Accidental Death and Disability: The Neglected Disease of Modern Society" (National Academy of Sciences), 82
additive manufacturing, 225–226
Adner, Ron, 217–218
The Advancement of Learning (Bacon), 54–55
Age of Discovery, 55–56
AI. *See* artificial intelligence (AI)
AirPods, 202, 232
Alfred Knopf, 131–133, 134, 140–141
al-Rammah, Hasan, 26–28
Amazon, 2, 212
    Prime, 158
    Web Services, 173–176
ambiguity, adeptness in, 139, 141
American Research and Development Corporation (ARD), 117–118
American Restaurant Magazine, 145
Andrews, R. J., 74, 78
Anduril Industries, 231
Ann Frank: The Diary of a Young Girl, 131
anomalies, 99, 216–219
The Anxious Generation (Haidt), 214
Apple, 2, 118, 201–219, 232–233
    App Store, 208, 218, 233
    innovators at, 203–206
    iPad, 202, 232
    iPhone, 22, 201–219, 232–233
    iPod, 201, 202, 208
    Vision Pro, 232–233
    Watch, 202, 232
    App Store, 208, 218, 233
ARD. *See* American Research and Development Corporation (ARD)
Aristotle, 52, 55
artificial intelligence (AI), 2, 100–101, 212, 226–227
assembly line, 91–94
astronomy, 51
AT&T, 8, 105, 111–112
automobiles and automobile industry, 85–101
    autonomous, 227–228
    conflicts over roads and, 85–86, 215
    democratization and, 96–98
    electric vehicles, 197–199
    Ford and, 87–98
    impact of, 99–101
    Model T, 91–94
    technological developments in, 87–89
    worker turnover in, 94–96
autonomous vehicles, 227–228
A&W Restaurant, 147
Aykroyd, Dan, 134–135

Bacon, Francis, 15, 42, 53–60, 64, 212
Bardeen, John, 103–104, 105–107, 108, 115, 231

Bedside Rounds (podcast), 71
Behind the Arches (Love), 152
Bell, Alexander Graham, 7–8, 111–112
Bell Telephone Laboratories, 2, 105–112
Bertholle, Louisette, 127–128, 140
Bessemer, Henry, 183
Bethlehem Steel, 179–187, 189–191, 193–199, 215
Beyond Meat, 233
Bezos, Jeff, 175, 212
Biringuccio, Vannoccio, 26, 30
Bitcoin, 230
BlackBerry, 22
Black Mirror (TV show), 227
blasting caps, 20
Blockbuster, 2, 157–158
Boehret, Katherine, 208–209, 216–217
Bohmer, Richard, 78
The Book of the Kinship of Three, 15
Book of the Master Who Embraces Simplicity (Ge), 15
Books of Fire for the Burning of Enemies (Marcus Graecus), 26
Boston Dynamics, 235
Boughton, Jill, 170, 173
Bower, Joe, 4–5, 6, 182, 216
Boyle, Robert, 53, 58–60, 64, 147, 178
"Boys with the Bus," 136–137, 138
Brattain, Walter, 103–104, 105–107, 115–116, 231
Brooks, Paul, 130–131
Brown, Brené, 195
Brownie box camera, 7
Bryan, Ford R., 87, 97–98
bubonic plague, 33
Bugnard, Max, 127, 128
Busch, Christian, 177
business models
    additive manufacturing and, 225–226
    Amazon Web Services, 173–176
    future of, 207, 227, 229, 230, 234
    McDonald's, 147–156, 234
    Netflix, 156–158
    Pampers, 160

Buxton, Bill, 203, 205–206
Byzantine Empire, 13–14, 19–20

Cadillac, 89
Calvel, Raymond, 140
Campbell, Joseph, 127
cannons, 17, 19–20, 21
carbon dioxide removal technologies, 229
cardiopulmonary resuscitation, 82–83
Carnegie, Andrew, 183–184, 186, 193–194, 235
Casey, Bob, 2
Catholic Church, 197–198
    Gutenberg's Bible for, 37–42
    printing press and, 31, 32, 45–46
Le Cercle des Gourmettes, 127–128
change
    acceleration of, 11, 208–212
    adoption of, 62–63
    future of disruption and, 221–238
    loss avoidance and, 47
    pace of in innovation *vs.* people's lives, 5–6, 234
    power and, 47, 50
    status quo effect and, 62–64
"The Charge of the Light Brigade" (Tennyson), 70
Charles VII, 18–19
ChatGPT, 100–101, 212, 226
Chemcor, 209–210
Child, Julia, 123–141, 147, 212, 230
    behaviors of disrupters and, 138–141
    Great French Bread Experiment of, 140–141
    hero's journey of, 126–133
    impact of, 136–138
    TV and, 133–135
Child, Paul, 123–124, 129, 133
China
    gunpowder developed in, 15–17
    Pampers and, 169–173
    printing press and, 43

Christensen, Clay, 3–8, 23, 63, 237–238
  on anomalies, 99, 216–217
  on competing against nonconsumption, 119
  on health care, 78
  on innovators' behaviors, 138–141
  Intel and, 222–223
  on iPhone, 215, 216–217, 219
  on Pampers, 171
  on steel, 179–182, 190–191
Church of England, 52
Church of Jesus Christ of Latter-Day Saints, 216
CK718 transistor, 112
Claiborne, Craig, 132–133
Clark, Kim, 4
Classified Essentials of the Mysterious Tao of the True Origin of Things, 16–17
cleantech, 229
cloud computing, 174–176
cobots, 234–235
coin making, 32, 33
collaboration, 139, 140, 237
Collins, Jim, 191
Columbus, Christopher, 55–56
common language, benefits of, 27, 223–224
Competing Against Luck (Christensen, Hall, Dillon, & Duncan), 170, 171–173
Constantine I, 13–14
Constantinople, 13–14, 19–20
Cook, Tim, 202
Copernicus, Nicolaus, 51
Corning, 209–211
La Couronne restaurant, 124–125
coxcomb chart. *See* polar area charts
creative destruction, 197–198
Crest, 159
Crimean War, 68–79
Crisco, 159
cruise control, 227–228
Crystal Fire (Riordan & Hoddeson), 106
curiosity, 139, 140

customers
  creating value for, 148–150
  delivering value to, 150–153
  obsession with, 139, 140–141

Dairy Queen, 147
Dardanelles gun, 19–20, 21
DARPA. *See* Defense Advanced Research Projects Agency (DARPA)
data, Nightingale and, 67–68, 73–76
data storytelling, 78
Dedu, Horace, 212
Defense Advanced Research Projects Agency (DARPA), 231
De la pirotechnia (Biringuccio), 26
democratization, 2, 96–98, 121, 133, 136–138
de Santillana, Dorothy, 129, 130
Detroit Journal, 95
DeVoto, Avis, 129, 131
DeVoto, Bernard, 129
Dexter, George, 146
Diffusion of Innovation (Rodgers), 62
Digital Equipment Corporation, 2, 118
digital health solutions, 236
digital imaging, 46–47
Dillon, Karen, 170
disk drive industry, 4
disruption
  acceleration of, 11, 208–212
  behaviors driving, 138–141
  case studies on, 12
  definition of, 7
  dismissals of, 21–23, 46–47, 157–158
  downsides from, 45–47, 100–101, 197–198, 213–215
  health care, 78–79
  industry change from, 2, 8
  innovation as, 1–3
  intersections and, 42–44, 93–94, 206–208
  Julia Child and, 131–133, 138–141
  McDonald's, 143–155
  media, 136–138

disruption (*Continued*)
  model of disruptive innovation, 4–5
  the Model T, 89–98
  overshooting and, 5–6
  Pampers, 159–161, 163–173
  progress and, 2
  recipe for. *See* behaviors driving
  responding to, 218–219
  responsibility of the many in, 63–65
  scientific revolution, 53–60
  status quo and, 61–63
  steel industry, 179–182, 187–191,
  system-level approach in, 155–156
  theory of, 3–9, 215–219
  transistors, 111, 119–121
  unanswered questions about, 9–12
  as universal good, 11, 213–215
  who does it?, 203–206
The Disruption Dilemma (Gans), 219
distributed ledger technologies, 230
Doerr, John, 175
drones, 231
Duhamel, Albert, 133
Duncan, David S., 170
Duncan, Robert, 162, 163–164, 176–177
Dyer, Davis, 162, 167, 168
Dyer, Jeff, 138–139

Eastman, George, 7
Eastman Kodak, 2, 7
  digital imaging and, 46–47
Eat, Sleep, Innovate (Anthony, Cobban, Painchaud, & Parker), 61
ecosystems, 206–208, 217–218, 232–233
Edgar Thomson Steel Works, 183–184
Edison, Thomas, 87, 183
Edison Illuminating Company, 87
Edward III, 18
Edward VI, 49–50, 52–53, 63, 214–215
Eisenhower, Dwight, 187
empowerment, 139, 141
experimentation, 147, 178
  Bacon and, 56, 57, 59
  at Bell Labs, 108
  Boyle and, 60
  gunpowder and, 25
  at McDonald's, 155
  Nightingale and, 80–81
  Pampers and, 164–165, 169–173, 212
  replication and, 60
ExxonMobil, 2

Facebook, 46–47, 218
Fairchild Camera and Instrument, 115
Fairchild Semiconductor, 115–116, 117
Farr, William, 73–76
financing, venture capital in, 116–118
FingerWorks, 204–205
firecrackers, 17
Fischbacher, Simone Beck, 127–128, 129–131, 133, 140–141
flivverboobs, 86, 100, 215
Ford, Henry, 2, 85–101, 105, 199
  death of, 97–98
  democratization and, 96–98
  hero's journey of, 126
  mass production and, 91–94
  paradoxes and, 98–99
  system-level approach by, 155–156
  vision of, 89–91
  worker turnover and, 94–96
Ford Motor Company, 91–94
Forging America (Hilliard & Venditta, eds.), 194, 196
The Founder (movie), 147
Foxconn, 213–214
Franchise Realty Corporation, 154
franchises
  McDonald's, 145–147
  short- *vs.* long-term orientation in, 150–153
Freedom House, 81–84
The French Chef (TV show), 133–135
Friendly, Michael, 68, 77
Frisch's restaurant, 151–152
Füssel, Stephan, 34
Fust, Johann, 30, 38–39, 40–41
Galilei, Galileo, 53

Gamble, James A., 161
Gamble, James Noble, 161–162
Gans, Joshua, 219
Gardiner, Bryan, 210, 211
Gates, Bill, 229
General Motors, 2, 88, 89
General Register Office, 73, 74
generative artificial intelligence
    (gen AI), 100–101, 226–227
Germanium Products Corporation, 112
ghosts, organizational, 199
    Bethlehem Steel, 193–197
Gilbert, Clark, 216
Glackin, George, 169
Godin, Benoît, 50–51, 53, 63–64, 65
Golden Sleep campaign, 172
Goldman Sachs, 172
Good to Great (Collins), 191
Google, 137, 158, 232
    Android, 22
Google Glass, 232
Gorilla Glass, 210–211
Gou, Terry, 214
Goulait, Dave, 169
Grace, Eugene Gifford, 185–187,
    193–194, 196
Grant, Adam, 79
Gregersen, Hal, 138–139
Grinding It Out (Kroc with Anderson),
    144
Groen, Lou, 151–152
Grossman, Jerome, 78
Grove, Andy, 8, 117–118, 222–223
Guerry, André-Michel, 68, 73
gunpowder, 14–28
Gunpowder (Charles River Editors),
    16, 25
Gunpowder: Alchemy, Bombards &
    Pyrotechnics; The History of the
    Explosive That Changed the World
    (Kelly), 18–19
Gunpowder Plot, 54–55
Gutenberg, Johannes, 26, 29–42, 226
    Bible printed by, 37–42
    early years of, 32–34

    innovations utilized by, 34–36
    lone hero story and, 205
    mirrors by, 33–34
    successes and failures of, 39–42
The Gutenberg Revolution (Man), 31, 33,
    35–36, 41–43
Gutenberg! The Musical!, 30–31

Haidt, Jonathan, 214
Hall, Taddy, 170
Hamburger University, 153, 155
"Hand Tracking, Finger Identification,
    and Chordic Manipulation on a
    Multi-Touch Surface" (Westerman),
    203–204
Hastings, Reed, 157
Hayden, Stone & Co., 115
health care, 67–84, 226
    enabling a broader population in,
        79–81
    evidence-based, 72–79
    paramedics and, 81–84
    smart health and, 235–237
hearing aids, 111–112, 119
Heath, Chip, 27
Heath, Dan, 27
Henry Ford Company, 88
Henry V, 18
Herbert, Sidney, 70, 73, 74, 75
hero story
    of Child, 126–133
    of gunpowder, 23–28
    Gutenberg and, 41–42
The Hero with a Thousand Faces
    (Campbell), 127
Hilliard, Ardith, 194, 196
Hoddeson, Lillian, 106, 108–109, 112
Hoerni, Jean, 115
Hon Hai Precision Industry Company,
    Ltd., 213–214
Hooke, Robert, 53, 58–59
Houghton Mifflin, 129–131
How to Avoid a Climate Disaster
    (Gates), 229

Huggies, 168
Hughes, Nick, 119–120
Hundred Years' War, 18–19
Huy, Quy, 22
Hwang, Jason, 78

IBM, 4, 118
Ibuka, Masaru, 113–114
IDEA. *See* Industrial Development & Engineering Associates (IDEA)
identity, fear of losing, 195–197
iGesture Pad, 204–205
IG Farben, 162
Impossible Foods, 233
individualism, 10, 117
inductive reasoning, 57
Industrial Development & Engineering Associates (IDEA), 113
Industrial Revolution, 64, 182–183
Informal Records of Eminent Physicians (Tao), 15
information
  the Church's control of, 31, 32
  printing press and, 29–47
  spread of on gunpowder, 25–26
infrastructure, 96–97
Innosight, 61, 169, 216
innovation
  anomalies and, 99
  as corporate value, 61
  definition of, 27, 63–64
  as deviance, 49–51, 52–53
  diffusion of, 62
  jobs to be done and, 118–121
  long nose of, 205–206
  nonlinear paths to, 107–111
  organizational ghosts and, 194–197
  perseverance and, 11, 176–178, 212
  twisted paths to success and, 160–161
  venture capital for, 116–118
  vocabulary for discussing, 27
innovation, disruptive. *See* disruption
Innovation Contested (Godin), 50–51, 53, 63–64, 65

The Innovator's Dilemma (Christensen), 3–4, 191, 219
The Innovator's DNA (Christensen, Dyer, & Gregersen), 138–141
The Innovator's Prescription (Christensen, Grossman, & Hwang), 78
The Innovator's Solution (Christensen & Raynor), 23
Instagram, 138, 218
Instauration Magna (Bacon), 55
Intel, 8, 116–118, 222–223
The International Jew (Ford), 98–99
International Steel Group, 190
intersections, 235
  automobile assembly line and, 93–94
  collaboration and, 139
  predictable unpredictability and, 206–208
  printing press and, 42–44
iPad, 202, 232
iPhone, 2, 201–219
  acceleration of innovation and, 208–212
  app ecosystem for, 206–208, 217–218, 232–233
  creation of, 201–202
  disruption theory modernization and, 215–219
  innovators behind, 203–206
  randomness and, 206–208
  screen of, 208–211
  shadow cast by, 213–215
iPod, 201, 202, 208
Isaacson, Walter, 210–211, 219
Isabella of Spain, 55–56
iTunes, 208
I've Been Reading (TV show), 133
Iverson, Ken, 181, 187–189, 234
Ives Washburn, 127–128, 129
Ivory soap, 161–162, 176

Jassy, Andy, 174–175
jaywalkers, 86, 100, 215
Jewkes, John, 206

Joan of Arc, 18–19, 44
Jobs, Steve, 118, 201–202, 206–207, 208, 210–211, 219
jobs to be done, 119
    Pampers and, 169–173
Johnson, Steven, 44
Johnson & Johnson, 163
Jones, Judy, 131–133, 140–141
Jope, Alan, 221
J. P. Morgan, 184
Julia's Kitchen Wisdom (Child), 135
Justinian II, 13–14

kaidangku, 169–173
Kallasvuo, Olli-Pekka, 22–23
Keaton, Michael, 147
Kelly, Jack, 18, 24
Kempczinski, Chris, 152
Kenagy, John, 78
Kennedy, John F., 132
Kepler, Johannes, 53
Kimberly-Clark, 168
King, Anthony, 30–31
Kleiner Perkins Caufield & Byers, 118, 175
knowledge
    perceived limits of, 55–56
    printing press and the spread of, 30–42, 44–47, 49–53
    spread of, about gunpowder, 25–26
Kroc, Ray, 97, 143–144, 148–150, 151–152, 153–155, 212. *See also* McDonald's

labor disputes, 186–187, 195
Lead and Disrupt (O'Reilly & Tushman), 175
Le Cordon Blue, 127
Lehigh Valley Live, 197, 198
Leland, Henry M., 89
Lenovo, 2
Lepore, Jill, 219
Levitt, Theodore, 96

Little Steel Massacre, 186
lone genius theory, 10, 105, 205–206
    gunpowder and, 24–28
    printing press and, 41–42
    responsibility of the many *vs.*, 63–65
long-term orientation, 150–153, 173–176
loss avoidance, 47
Love, John, 145, 152
luck, 176–177
Luckey, Palmer, 231
The Luck Factor (Wiseman), 177
Luther, Martin, 30, 42, 45

Man, John, 31, 33, 35–36, 41, 42–43
manufacturing, additive, 225–226
"Marketing Myopia" (Levitt), 96
Martin, Edmund, 190
Martino, June, 149
mass production, 91–94
Mastering the Art of French Cooking (Child, Bertholle, & Beck), 123–141
McCarthy, John, 212, 226
McDonald, Maurice, 145–148, 153–154. *See also* McDonald's
McDonald, Richard (Dick), 145–148, 153–154. *See also* McDonald's
McDonald's, 2, 143–158, 212
    architecture of, 145–146
    brief history of, 145–155
    business model of, 147–156, 234
    delivering value to customers, 150–153
    value capture by, 153–155
McDonald's: Beyond the Arches (Love), 145
McKinsey & Co., 174
meat, lab-grown, 100–101, 233–234
media, disruption of, 136–138
Mehmed II, 19–20
Merchant, Brian, 205, 213
Meta, 218, 232
Microsoft, 218
Mills, Victor, 159–161, 176–177
Minsky, Marvin, 212, 226
Mintzberg, Henry, 177–178

mixed reality, 232–233
Model K, 90
Model N, 2
Model T, 2, 85–101, 199
   brief history of, 87–98
Monitor Group, 216
monomyth, 126–127
Moore, Gordon, 115, 117–118
Morgan, J. P., 184
Morita, Akio, 113–114
"Mortality of the British Army, at Home, at Home and Abroad, and During the Russian War, as Compared with the Mortality of the Civil Population in England," 75
Mossberg, Walt, 208–209, 216–217
Motorola, 22, 206–207, 210
M-PESA, 118–121
Multimixer, 143–144

Nadella, Satya, 218
National Academy of Sciences, 82
Nespresso, 2
Nestlé, 2
Netflix, 156–158
New Atlantis (Bacon), 58
New Experiments Physico-mechanicall Touching the Spring of Air and Its Effects (Boyle), 59–60
new food, 233–234
Newton, Isaac, 53
The New World of English Words, 64
New York Times, 110
New York Times Magazine, 185
Nicholas of Cusa, 37, 39
Nightingale, Florence, 68–81
   hero's journey of, 126
Nobel, Alfred, 20
Nobel Prize, 103, 115
Nokia, 2, 21–23
nonconsumption, competing against, 118–121
Notes on Nursing: What It Is and What It Is Not (Nightingale), 80–81

Novum Organum (Bacon), 55
Noyce, Robert, 114–115, 117–118
Nucor Corporation, 181–182, 187–189, 190, 195, 223, 234
Nussbaum, David, 135–136
Nvidia, 218

Office of Strategic Services (OSS), 123–124
Ofoto, 46–47
"Of Strategies, Deliberate and Emergent" (Mintzberg & Waters), 177–178
Olds, Ransom E., 88, 93, 187
Oldsmobile, 88, 89
The One Device: The Secret History of the iPhone (Merchant), 205
OpenAI, 100–101, 212, 226
Orban, 19–20, 21
O'Reilly, Charles, 175
organizational design, 224
origin stories, 157
Orton, William, 7
OSS. *See* Office of Strategic Services (OSS)
Otto, Nikolaus, 88
Ottoman Empire, 14
overshooting, 5–6, 116–118

Pampers, 159–178, 212
   brief history of, 161–169
   Change 'N Go, 170, 171–173
   in China, 168–173
   launch of, 166–167
   market for, 166–169
   price point for, 165–169
   revenues from, 160
paradoxes, 37, 42
   Ford and, 98–99
Parke-Davis, 163
Parker, Andy, 15
Pepper, John, 168–169
perseverance, 11, 176–178, 212
photography, 7, 46–47

Pius II, Pope, 40
Plato, 56
polar area charts, 67–68, 72–79
predictability and unpredictability, 10, 206–208
Prescriptions Worth a Thousand Gold (Sun), 16
printing press, 26, 29–47
  brief history of, 30–42
  the Church and, 37–42
  downsides of, 45–47, 197–198
  impact of, 42–47
  innovations contributing to, 34–36, 205
  scientific revolution and, 49–50
  system-level approach and, 156
  "A Proclamation Against Those That Doeth Innouate" (Edward VI), 49–50, 52–53, 214–215
Procter, Harley, 161–162
Procter, William, 161
Procter & Gamble, 159–178
  disposable diapers and, 163–173
  Exploratory Development Division, 159–161
  history of, 161–163
  innovation incubation at, 169–173
  perseverance at, 176–177
progress, 2, 64
proof of concept
  of the printing press, 36–37
  transistors, 108–109
The Prosperity Paradox (Christensen, Dillion, & Ojomo), 6
prototypes and prototyping, 46, 146–147, 188
Public Health Act of 1875 (Britain), 78
Pyroceram, 209

quadricycle, 88–89
Quest 3, 232

radios, transistor, 112–114
rational decisions, 192–193

Raynor, Michael, 23
Raytheon, 112
Razr V3, 210
religion, science and, 56–57, 60
renewable energy, 2, 229
REO Motor Cars, 187
Research in Motion, 22
resistive touchscreens, 204–205
resource allocation, 6, 198–199
  Bethlehem Steel and, 179–182, 189–191, 192–197
  long-term orientation and, 173–176
Rickey, Branch, 177
Riordan, Michael, 106, 108–109, 112
Rising Tide (Dyer, Dalzell, & Olegario), 162, 167, 168
Robinson, Jackie, 177
robotics, 234–235
Rock, Arthur, 115, 118
Rodgers, Everett, 62
ROKR, 206–207
Ronn, Karl, 60
Royal Commission, 72, 73, 78
Royal Radar Establishment, 203

Safaricom, 120
saltpeter, 15, 16, 24–25. *See also* gunpowder
sanitation, 71, 72–79
Sasson, Steve, 46
Saturday Night Live (TV show), 134–135
scale economics, 105, 168, 196
The Sceptical Chymist: or Chymico-Physical Doubts & Paradoxes (Boyle), 60
Schöffer, Peter, 30, 41
Schwab, Charles Michael, 183–187
science
  printing press and, 45
  truth and, 31
scientific method, 178. *See also* scientific revolution.
scientific revolution, 49–65
The Scientific Revolution (Shapin), 55

Scott Paper, 160
Sears, 2
semiconductors, 107–108, 114–116
Sequoia Capital, 118
The Serendipity Mindset (Busch), 177
Shakespeare, William, 14, 42, 58
Shapin, Steven, 55, 56, 59
Sharp, 114
Shell, 2
Shockley, William, 103–110, 114–116, 231
   hero's journey of, 126
Shrive, John, 109
smart health, 235–237
smartphones, 2, 214
   Nokia and, 21–23
SMS Schloemann-Siemag, 188
social media, 136–138, 214
Society of Automotive Engineers, 228
Sonneborn, Harry J., 154, 156
Sonotone, 112
Sony, 113–114
Sorenson, Charles E., 90
The Sources of Invention (Jewkes, Sawers, & Stillerman), 206
Speedee Service System, 145, 146–147, 148–150
Spitz, Bob, 126, 133
standardization
   in health care, 78, 84
   at McDonald's, 145, 148–150
   printing press and, 37–38
Statistical Society, 73
status quo, 47, 50, 61–63, 214–215
status quo effect, 62–64
steel industry, 179–199
   history of, 182–183
   integrated mills in, 180–182
   minimills in, 181–182, 187–189, 190–191, 192–197, 198, 222–223
   thin-slab casting, 188–189, 226
St. Louis Star, 86
Stookey, Donald, 209, 211
Strategic Management Journal, 177–178
streaming, 157–158
St Thomas' Hospital, 79

success, maintaining, 168–169
Sull, Don, 61
superconductivity, 115
Sweeney, Bud, 152
Switch (Heath & Heath), 27
systems approach, 155–156

Tastee-Freez, 154
Tecklenburg, Harry, 165–166, 176–177
Tedlow, Richard, 23, 98, 99
Teetor, Ralph, 227–228
telegraphy, 7–8, 72
Tennyson, Alfred, Lord, 70
Tesla, 199
Texas Instruments (TI), 112–113
Theodosian Walls, Constantinople, 13–14
Theodosius II, 13–14
3D Systems, 225
TI. *See* Texas Instruments (TI)
Tide, 159, 162, 176
TikTok, 138
Time, 168
Today, 133
Tokyo Tsushin Kogyo, 113–114
Torricellian apparatus, 60
Toshiba, 114
touchscreens, 203–205
Traitorous Eight, 114–115
transistors, 2, 8, 103–121
   brief history of, 105–116
   hearing aids and, 111–112
   impact of, 110–111, 118–121
   Intel, venture capital, and, 116–118
   nonlinear paths to, 107–111
   transistor radios and, 112–114
trauma, historical, 194–195
Turner, Fred, 149–150, 151, 156
Turner Broadcasting, 137
Tushman, Michael, 175

Ukraine war, 231
Ultra Pampers, 168

uncertainty, 88–89
Unilever, 221
United States Steel (U.S. Steel), 184

vacuum tubes, 105, 107, 110–111, 119
value
  capturing for yourself, 153–155
  creating for customers, 148–150
  delivering to customers, 150–153
  innovation and creation of, 27
Venditta, David, 194, 196
venture capital, 116–118
Verdon, René, 132
vertical integration, 93–94
Victoria I, 69, 72
virtual reality, 232–233
Vision Pro, 232
Vodafone, 119–120
Volkswagen, 2
Vulcraft, 187–189

Wall Street Journal, 96, 189, 208–209
Waters, James A., 177–178
Wedell-Wedellsborg, Thomas, 191
Weeks, Wendell, 210–211
"A Well-Deserved Rehabilitation of Johann Gutenbergs, as Amply Attested in Surviving Documents" (Köhler), 41–42
Westerman, Wayne, 203–204
Western Electric, 106–107

Western Union, 7–8
WhatsApp, 218
What's Cooking in France, 127–128
"What the Gospel of Innovation Gets Wrong" (Lepore), 219
Where Good Ideas Come From (Johnson), 44
The Wide Lens (Adner), 217–218
"Will Disruptive Innovation Cure Health Care?" (Christensen, Bohmer, & Kenagy), 78
Winning the Right Game (Adner), 217–218
Winton, Alexander, 88–89
Winton Motor Carriage Company, 88–89
The Wire (TV show), 52–53
Wiseman, Richard, 177
Wittgenstein, Ludwig, 27
woodcut printing, 31–32
worker turnover, 94–96, 145
working conditions, 95–96, 214
World War I, 185
World War II, 107–108, 185–187, 231

Xerox Corporation, 115

YouTube, 137, 138

Zuckerberg, Mark, 46–47

Acknowledgments

This was the hardest part of the book to write.

The beginning was easy.

I'm ever grateful for the love and support of friends and family, particularly my wife, Joanne, and our children, Charlie, Holly, Harry, and Teddy. I've shared pieces of the book with them during its development and have enjoyed shaping the ideas with them.

Charlie first appeared in my 2009 book *The Silver Lining* for a line he once said to his Montessori teacher Ms. Wendy: "The dark clouds are rolling in." Watching that precious child turn into a poised young man has been a journey of wonder. Holly first appeared in *The Little Black Book of Innovation*; she was playing with a Dora the Explorer laptop, which was surely powered by an offspring of the transistor from Bell Labs. Harry has firsthand experience with four of the echo disruptions in *Epic Disruption's* conclusion and seems on track to be quite the disrupter in his own right. Teddy makes his first appearance in this book (in a footnote about broccoli). I'm excited to see what he does to deserve a feature story in the next one. And Joanne, I suspect we'll never forget the moment I blew Julia's chocolate mousse. Thank you for everything.

This is my ninth dance with Harvard Business Review Press. It's always fun. Kevin, thanks for the suggestion that got this started and the support along the way. Melinda, thanks for our own "lunch that changed everything." Victoria, thanks for getting it over the line. Patricia, all I can say is "wow." You did just a beautiful job with your copy edits. Adi, thanks for being a friend. Julia, Alexandra, and the rest of the team, thanks for the great support that I know will come after these words are written.

Four friends provided particularly valuable feedback on the manuscript: Thomas Wedell-Wedellsborg (my favorite name to write), Karl Ronn, Clark Gilbert, and Roger Martin. As I tell my students, I welcome all feedback and listen to some of it. Kim Clark and Joe Bower were kind enough to share memories of Clay's original research.

. . .

It gets harder.

Clay.

Writing this book led to constant reminders of our close-to-twenty-year relationship. I felt constant gratitude for the two years I got to spend leading his research team from 2001 to 2003 and for the more than fifteen years of collaboration that followed during my time at Innosight.

In December 2019, I was to pay a visit to his home, but his wife, Christine, sent me an email saying that he was too sick to take visitors. On January 23, 2020, I was in a taxi on my way to Logan Airport. I received a text from Mark Johnson, who cofounded Innosight with Clay in 2000. Clay had died.

It's hard for me to find the words to describe how I felt at that moment. I had seen Clay at his best moments, and I had seen him at some of his worst. He was human. He could get frustrated. But no matter how tired he was, he always tried so hard to be kind, caring, helpful. I watched him, learned from him, sought to develop my own approach, my own style, my own thinking.

Clay was a deeply religious man, not shy about sharing his faith in meetings or when he spoke. I am not of his church, but I attended breakfasts that started in prayer, heard stories about his mission and his children's missions, heard him use religious parables to explain business concepts.

When Clay died, there was an outpouring of collective grief. The thread that connected almost every contribution was a reflection on the man himself. His humility. His graciousness. His curiosity. His willingness to learn. I've never met a better teacher; I've never seen a kinder man.

Clay was an intellectual giant, no doubt. But his gift to the world was more than models, frameworks, case studies, stories, and books. His greatest gift was a template for how to live a humble, intellectually curious life. He believed ideas matter. He believed ideas could in fact change the world. At its core, innovation is an optimistic activity. It is grounded in the belief that tomorrow can be better than today. That tomorrow should be better than today. That today's ambiguity creates tomorrow's opportunities.

I think of Clay's guidance often. When I stray from it—and I have strayed more frequently than I would like to admit—I hear his voice in my head. The voice never takes a judgmental tone. Rather, I hear Clay asking me to think of why I did the things I did, what I expected, and what I learned. I strive to live up to his legacy. I'm sure I'll fall short, but the journey is worth it.

I am drafting this while staring out of an airplane window on a flight from Boston to Las Vegas. It is a brutally cold morning. Above the clouds, the sun shines in my eyes as I remember my friend. The world is better because he was in it. I hope these words have captured the essence of him appropriately. I miss him. A lot.

. . .

And now, the hardest part.

I tell students in my class Leading Disruptive Change that they will get the most from the class if they bring their whole self to it, if they are vulnerable and willing to explore things that lurk in the shadow. So, here goes.

On December 10, 2024, I was driving to a doctor's appointment, chatting with my former Innosight colleague Andy Parker. My brother Peter called. I didn't answer. He called again. I took his call. He told me that our other brother, Michael, was dead. He was forty-five years old.

I am writing these words eight weeks after that call. I am still in shock.

Michael ran a veterinary practice. At his memorial, countless clients and members of his staff spoke of his kindness, his generosity, his desire to help others. He was a very, very different person from Clay but a kindred spirit in the overwhelming desire to put other people first.

I have found peace and solace working through the words in this manuscript. It makes me both happy and sad to think about how excited Michael would be to read it. We would both laugh at the Mr. Brightside reference in the Julia Child chapter. He would have appreciated being called out in a footnote about *The Simpsons*.

It was always easy with my brother.

I miss you so much, Mike. This one, above all else, is for you. May these words make it through the universe and bring you a smile wherever you are.

About the Author

Scott D. Anthony is a Clinical Professor of Strategy at the Tuck School of Business at Dartmouth College, where his research and teaching focus on the adaptive challenges of disruptive change. *Epic Disruptions* is Scott's ninth book. His previous books are *Seeing What's Next*, with coauthors Clayton Christensen and Erik Roth; *The Innovator's Guide to Growth*, with coauthors Mark Johnson, Joseph Sinfield, and Elizabeth Altman; *The Silver Lining*; *The Little Black Book of Innovation*; *Building a Growth Factory*, with coauthor David Duncan; *The First Mile*; *Dual Transformation*, with coauthors Clark Gilbert and Mark Johnson; and *Eat, Sleep, Innovate*, with coauthors Paul Cobban, Natalie Painchaud, and Andy Parker.

Scott previously spent more than twenty years at Innosight, a growth strategy consultancy cofounded by Harvard Business School Professor Clayton Christensen. Scott served as Innosight's elected managing partner from 2012 to 2018, a period where Innosight tripled its revenues and expanded internationally. In 2017, Huron Consulting acquired Innosight for $100 million.

Scott has lived in the United Kingdom (1997–1998) and Singapore (2010–2022), held board roles at public and private companies, given keynote address on six continents, and worked with CEOs at numerous global organizations. Thinkers50 named him the world's ninth-most influential thinker in 2023. He has been nominated for the Thinkers50 Innovation Award three times and won the award in 2017.

Scott received a BA in economics from Dartmouth College, an MBA from Harvard Business School, and an Executive Master in Change from INSEAD.

He and his wife, Joanne, are the proud parents of four children.